Pallavi Sharma
Vivek Bajpai
Pankaj Kumar Jain

Microbial Ecology

Pallavi Sharma
Vivek Bajpai
Pankaj Kumar Jain

Microbial Ecology

Study of Microbial Applications

LAP LAMBERT Academic Publishing

Impressum/Imprint (nur für Deutschland/only for Germany)
Bibliografische Information der Deutschen Nationalbibliothek: Die Deutsche Nationalbibliothek verzeichnet diese Publikation in der Deutschen Nationalbibliografie; detaillierte bibliografische Daten sind im Internet über http://dnb.d-nb.de abrufbar.
Alle in diesem Buch genannten Marken und Produktnamen unterliegen warenzeichen-, marken- oder patentrechtlichem Schutz bzw. sind Warenzeichen oder eingetragene Warenzeichen der jeweiligen Inhaber. Die Wiedergabe von Marken, Produktnamen, Gebrauchsnamen, Handelsnamen, Warenbezeichnungen u.s.w. in diesem Werk berechtigt auch ohne besondere Kennzeichnung nicht zu der Annahme, dass solche Namen im Sinne der Warenzeichen- und Markenschutzgesetzgebung als frei zu betrachten wären und daher von jedermann benutzt werden dürften.

Coverbild: www.ingimage.com

Verlag: LAP LAMBERT Academic Publishing GmbH & Co. KG
Heinrich-Böcking-Str. 6-8, 66121 Saarbrücken, Deutschland
Telefon +49 681 3720-310, Telefax +49 681 3720-3109
Email: info@lap-publishing.com

Herstellung in Deutschland:
Schaltungsdienst Lange o.H.G., Berlin
Books on Demand GmbH, Norderstedt
Reha GmbH, Saarbrücken
Amazon Distribution GmbH, Leipzig
ISBN: 978-3-8465-3830-2

Imprint (only for USA, GB)
Bibliographic information published by the Deutsche Nationalbibliothek: The Deutsche Nationalbibliothek lists this publication in the Deutsche Nationalbibliografie; detailed bibliographic data are available in the Internet at http://dnb.d-nb.de.
Any brand names and product names mentioned in this book are subject to trademark, brand or patent protection and are trademarks or registered trademarks of their respective holders. The use of brand names, product names, common names, trade names, product descriptions etc. even without a particular marking in this works is in no way to be construed to mean that such names may be regarded as unrestricted in respect of trademark and brand protection legislation and could thus be used by anyone.

Cover image: www.ingimage.com

Publisher: LAP LAMBERT Academic Publishing GmbH & Co. KG
Heinrich-Böcking-Str. 6-8, 66121 Saarbrücken, Germany
Phone +49 681 3720-310, Fax +49 681 3720-3109
Email: info@lap-publishing.com

Printed in the U.S.A.
Printed in the U.K. by (see last page)
ISBN: 978-3-8465-3830-2

MICROBIAL ECOLOGY

Edited by

PALLAVI SHARMA

M.Sc., Ph.D. (Per.)
Department of Microbiology
Bundelkhand University
Jhansi-284128
(U.P), India

Dr. VIVEK BAJPAI

M.Sc., Ph.D.
Assistant Professor of Microbiology
Department of Microbiology
MITS University
Lakshmangarh, Sikar-332311
Rajasthan, *INDIA*

Dr. PANKAJ KUMAR JAIN

M.Sc., Ph.D.
Assistant Professor of Microbiology
Department of Microbiology
MITS University
Lakshmangarh, Sikar-332311
Rajasthan, *INDIA*

LAP LAMBERT ACADEMIC PUBLISHING AG & CO. KG, DUDWELLER LANDSTR, GERMANY

Acknowledgement

At the beginning of any task, it is difficult to imagine final shape, but as many thoughts, new ideas and various pairs of beautiful hands, conglomerate, every task reaches its destination. As *"Srimmad Bhagwat Geeta"* says *"Every beginning has its end"*, this task is to its end and it is real time to recherish those hard earned moments. It is also a time to give thoughts to those persons, who made one way or other and acknowledge their wisdom, benevolence and patience.

It gives me pleasure in expressing my sincere gratitude to *Prof. Shakti Baijal (Dean-FASC)* for their inspiring guidance, constant encouragement and above all for her wonderful nature. Mam, from you I have learnt not only the importance of growing into a good academician but more importantly as a good human being.

I am highly obliged to *all my co-authors* for their scientific guidance and excellent support they provided, without which this work would not have been possible.

I would like to express my sense of Respect to my parents, *Smt. Divya Lekha Bajpai* and *Sri. Jamuna Prasad Bajpai* for their loving blessing and bearing the work tension with me. My most sincere and special thanks would go to my bhai *Er. Mayank Bajpai* and bhabhi *Mrs. Vandana Bajpai* for their special care and love. Their contribution had been of immense importance to complete the work.

I wish to put on record my heartfelt emotions to *Mast. Yash, Mast. Amogh and Baby Pihu* for their innocent wishes especially during the last crucial days of this work.

Lastly, I thank Almighty, The Great for giving me the strength, patience and courage to carry out this uphill task. Without His blessings, nothing would have been possible.

Pallavi Sharma
Dr. Vivek Bajpai
Dr. Pankaj Kumar Jain
Mody Institute of Technology and Science
Lakshmangarh-SIKAR (Rajasthan)

LAP LAMBERT ACADEMIC PUBLISHING AG & CO. KG, DUDWELLER LANDSTR, GERMANY

PREFACE

Microbial ecology is the relationship of microorganisms with one another and with their environment. Microorganisms are present in virtually all of our planet's environments, including some of the most extreme, from acidic lakes to the deepest ocean, and from frozen environments to hydrothermal vents.

Microbes, especially bacteria, often engage in symbiotic relationships with other organisms, and these relationships affect the ecosystem. One example of these fundamental symbioses is chloroplasts, which allow eukaryotes to conduct photosynthesis. Similarly Soil microorganisms are very important as almost every chemical transformation taking place in soil involves active contributions from soil microorganisms. In particular, they play an active role in soil fertility as a result of their involvement in the cycle of nutrients like carbon and nitrogen, which are required for plant growth. For example, soil microorganisms are responsible for the decomposition of the organic matter entering the soil (e.g. plant litter) and therefore in the recycling of nutrients in soil. For example *Azospirillum* induces the proliferation of plant root hairs which can result in improved nutrient uptake.

Pallavi Sharma
Dr. Vivek Bajpai
Dr. Pankaj Kumar Jain
MITS(Deemed University),
Lakshmangarh-SIKAR (Raj.)INDIA

LAP LAMBERT ACADEMIC PUBLISHING AG & CO. KG, DUDWELLER LANDSTR, GERMANY

Contents

LAP LAMBERT ACADEMIC PUBLISHING AG & CO. KG, DUDWELLER LANDSTR, GERMANY

CHAPTER-1

MULTIDISCIPLINARY NATURE OF ENVIRONMENTAL SCIENCE AND

MICROBIAL APPLICATIONS

Awanish Kumar and Pallavi Sharma*

Institute of Parasitology, McGill University, Macdonald Campus,

Ste-Anne-de-Bellevue, Quebec- H9X 3V9, Canada

*Department of Microbiology, Bundelkhand University, Jhansi 284128-India

ABSTRACT

Concerns of environmental studies especially with reference to microbes have been identified as important area of biology where background information is essential for a better understanding of our environment. Useful applications of microbes, study of environment and their correlation are valuable for living organism. Nature of environmental science is multidisciplinary and study of microbial application has a very lengthy account. It stresses on a balanced view of issues that affect our daily lives. These issues are related to the conflict between existing 'development' strategies and the need for 'environmental conservation. Microbial application in environment indicate towards the need to conserve biodiversity, the need to lead more sustainable lifestyles and the need to use resources more equitably. There is a need to change the way in which we view our own environment by a practical approach based on observation and self learning. There is the need to create a concern for our environment (either natural or artificial/manmade) that will trigger pro-environmental action; including activities we can do in our daily life to protect it. This book chapter give you information about the environment and role of microbes in current scenario that will lead to a concern for your own environment. When you develop this concern, you will begin to act at your own level to protect the environment we all live in. The chapter summarizes multidisciplinary nature of environmental science and microbial applications.

LAP LAMBERT ACADEMIC PUBLISHING AG & CO. KG, DUDWELLER LANDSTR, GERMANY

INDEX

INTRODUCTION AND ELEMENTS OF ENVIRONMENTAL SCIENCE

Environmental science is the science of physical phenomena in the environment which include the studies of sources, reactions, transport, effect and fate of physical a biological species in the air, water and soil and the effect of from human activity upon these. It deals with each and every issue that affects the life of organism. It is essentially a multidisciplinary approach because it comprises various branches of studies like physics, chemistry, life science, medical science, agriculture, geology, sociology, anthropology, statistics etc. It is an applied science as it's seeks practical answers to making human civilization sustainable on the earth's finite resources.

Environment is constituted by the interacting systems of physical, biological and cultural elements. It is inter-related in various ways, individually as well as collectively. These elements which constitute it are described below:

(1) Physical elements: Physical elements are as space, landforms, water bodies, climate soils, rocks and minerals. They determine the variable character of the human habitat, its opportunities as well as limitations. Physical environment refers to geographical climate and weather or physical conditions wherein and individual lives. The human races are greatly influenced by the climate.

(2) Biological elements: Biological elements such as plants, animals, microorganisms and men constitute the biosphere. In this chapter application of microorganism is described, which is one major of the major biological element.

LAP LAMBERT ACADEMIC PUBLISHING AG & CO. KG, DUDWELLER LANDSTR, GERMANY

(3) Cultural elements: Cultural elements includes as economic, social and political elements which are essentially manmade features. Social Environment includes an individual's social, economic and political condition wherein he lives. The moral, cultural and emotional forces influence the life and nature of individual behaviour.

Our environment consists of following four segments and microbes are everywhere, and their presence invariably affects the environment that they are growing in.

(1) Atmosphere: The atmosphere sustains life on the earth. It saves it from the hostile environment of outer space. It absorbs most of the cosmic rays from outer space and a major portion of the electromagnetic radiation from the sun. It transmits only here ultraviolet, visible, near infrared radiation (300-2500 nm) and radio waves. (0.14-40 m) while filtering out tissue-damaging ultra violate waves below about 300 nm. The atmosphere is composed of nitrogen and oxygen, argon, carbon dioxide, and trace gases.

(2) Hydrosphere: The Hydrosphere comprises all types of water resources oceans, seas, lakes, rivers, streams, reservoir, polar icecaps, glaciers, and ground water. 97% of the earth's water supply is in the oceans. About 2% of the water resources is locked in the polar icecaps and glaciers. Only about 1% is available as fresh surface water-rivers, lakes streams, and ground water fit to be used for human consumption and other uses.

(3) Lithosphere: Lithosphere is the outer mantle of the solid earth. It consists of minerals occurring in the earth's crusts and the soil e.g. minerals, organic matter, air and water.

(4) Biosphere: Biosphere indicates the realm of living organisms and their interactions with environment, viz atmosphere, hydrosphere and lithosphere.

ENVIRONMENT STUDIES: IMPORTANCE AND SCOPE

Environment is not a single subject. It is an integration of several subjects that include both natural science and social Studies. An understanding of the environment requires that we know what makes up the environment, and what its limits are and why is a scientific study of the environment important. We live in a world in which natural resources are limited. Water, air, soil, minerals, oil, the products we get from forests, grasslands, oceans and from agriculture and livestock, are all a part of our life support systems. Without them, life itself would be impossible. We waste or pollute large amounts of nature's clean water; we create more and more material like plastic that we discard after a single use; and we waste colossal amounts of food, which is discarded as garbage. Manufacturing processes create solid waste that are discarded, as well as chemicals that flow out as liquid waste and pollute water, and gases that pollute the air. These accumulate in our environment, leading to a variety of diseases and other adverse environmental impacts. At present a great number of environment issues, have grown in size and complexity day by day, threatening the survival of mankind on earth. We study about these issues besides and effective suggestions in the Environment Studies. Environment studies have become significant for the following reasons:

1. Environment issues: An international importance

It has been well recognised that environment issues like global warming and ozone depletion, acid rain, marine pollution and biodiversity are not merely national issues but are global issues and hence must be tackled with international efforts and cooperation.

2. Problems cropped in the wake of development

Development, in its wake gave birth to Urbanization, Industrial Growth, Transportation Systems, Agriculture and Housing etc. However, it has become phased out in the developed world. The North, to cleanse their own environment has fact fully, managed to move 'dirty' factories of South. When the West developed, it did so perhaps in ignorance of the environmental impact of its activities.

3. Explosively increase in world's pollution

A population of over thousands of millions is growing every year in the world. Over 17 million people are added each year. Due to excessive increase of world there is a heavy pressure on the natural resources including land. Agricultural experts have recognized soils health problems like deficiency of micronutrients and organic matter, soil salinity and damage of soil structure.

4. Need for an alternative solution

LAP LAMBERT ACADEMIC PUBLISHING AG & CO. KG, DUDWELLER LANDSTR, GERMANY

It is essential, especially for developing countries to find alternative paths to an alternative goal. We need a goal as under: (a) A goal, which ultimately is the true goal of development an environmentally sound and sustainable development. (b) A goal common to all citizens of our earth. (c) A goal distant from the developing world in the manner it is from the over-consuming wasteful societies of the "developed" world.

5. Need to save humanity from extinction

It is incumbent upon us to save the humanity from extinction. Consequent to our activities constricting the environment and depleting the biosphere, in the name of development.

6. Need for wise planning and management

Our survival and sustenance depend. Resources withdraw, processing and use of the product have all to by synchronised with the ecological cycles in any plan of development our actions should be planned ecologically for the sustenance of the environment and development.

As we look around at the area in which we live, we see that our surroundings were originally a natural landscape such as a forest, a river, a mountain, a desert, or a combination of these elements. Most of us live in landscapes that have been heavily modified by human beings, in villages, towns or cities. But even those of us who live in cities get our food supply from surrounding villages and these in turn are dependent on natural landscapes such as forests, grasslands, rivers, seashores, for resources such as water for agriculture, fuel wood, fodder, and fish. Thus our daily lives are linked with our surroundings and inevitably affects them. We use water to drink and for other day-to-day activities. We breathe air, we use resources from which food is made and we depend on the community of living plants and animals which form a web of life, of which we are also a part. Everything around us forms our environment and our lives depend on keeping its vital systems as intact as possible. Our dependence on nature is so great that we cannot continue to live without protecting the earth's environmental resources. Thus most traditions refer to our environment as 'Mother Nature' and most traditional societies have learned that respecting nature is vital for their livelihoods. This has led to many cultural practices that helped traditional societies protect and preserve their natural resources.

Over the past 200 years however, modern societies began to believe that easy answers to the question of producing more resources could be provided by means of technological innovations. For example, though growing more food by using fertilizers and pesticides, developing better strains of domestic animals and crops, irrigating farmland through mega dams and developing industry, led to rapid economic growth, the ill effects of this type of development, led to environmental degradation. The industrial development and intensive agriculture that provides the goods for our increasingly consumer oriented society uses up large amounts of natural resources such as water, minerals, petroleum products, wood, etc. Non-renewable resources, such as minerals and oil are those which will be exhausted in the future if we continue to extract these without a thought for subsequent generations. For example, if the removal of timber and firewood from a forest is faster than the regrowth and regeneration of trees, it cannot replenish the supply. Losses of forest cover not only depletes the forest of its resources, such as timber and other non-wood products, but affect our water resources because an intact natural forest acts like a sponge which holds water and releases it slowly. Deforestation leads to floods in the monsoon and dry rivers once the rains are over. Such multiple effects on the environment resulting from routine human activities must be appreciated by each one of us, if it is to provide us with the resources we need in the long-term. Our natural resources can be compared with money in a bank. If we use it rapidly, the capital will be reduced to zero. On the other hand, if we use only the interest, it can sustain us over the longer term. This is called sustainable development.

The environment studies enlighten us, about the importance of protection and conservation of our indiscriminate release of pollution into the environment. To understand all the different aspects of our environment we need to understand biology, chemistry, physics, geography, resource management, economics and population issues. Thus the scope of environmental studies is extremely wide and covers some aspects of nearly every major discipline.

MICROBIAL APPLICATIONS IN VARIOS ENVIRONMENTAL COMPONENT

Application of microorganisms are in fixation of carbon dioxide and nitrogen, decomposition of organic wastes and residues, detoxification of pesticides, suppress plant diseases and soil-borne pathogens, enhance nutrient cycling, and produce bioactive compounds such as vitamins, hormones and enzymes (Figure 1). Microbes maintain and improve human health, are economically and spiritually beneficial to both producers and consumers, actively preserve and protect the environment. As discussed earlier microbes are everywhere in the biosphere, the effects of microorganisms on their environment can be beneficial or harmful or in apparent with regard to human measure or observation. Since a good part of this text concerns with a discussion of the beneficial activities and exploitations of microorganisms as they relate to environment. The most significant effect of the microorganisms on earth is their ability to recycle the primary elements that make up all living systems, especially carbon (C), oxygen (O) and nitrogen (N). These elements occur in different molecular forms that must be shared among all types of life. Different forms of carbon and nitrogen are needed as nutrients by different types of organisms. The diversity of metabolism that exists in the microbes ensures that these elements will be available in their proper form for every type of life. Photosynthetic organisms take up CO_2 in the atmosphere and convert it to organic material. The process is also called CO_2 fixation, and it accounts for a very large portion of organic carbon available for synthesis of cell material. The unicellular organisms like algae and cyanobacteria account for nearly half of the primary production on the planet. Oxygenic photosynthesis occurs in plants, algae and cyanobacteria. It is the type of photosynthesis that results in the production of O_2 in the atmosphere. At least 50 percent of the O_2 on earth is produced by photosynthetic microorganisms (algae and cyanobacteria), and for at least a billion years before plants evolved, microbes were the only organisms producing O_2 on earth. O_2 is required by many types of organisms, including animals, in their respiratory processes. Biodegradation results in the breakdown of complex organic materials to forms of carbon that can be used by other organisms. These organic compounds are degraded by microbe. Through the metabolic processes of fermentation and respiration, organic molecules are eventually broken down to CO_2 which is returned to the atmosphere. Nitrogen fixation is a process found only in some bacteria which removes N_2 from the atmosphere and converts it to ammonia for use by plants and animals. Nitrogen fixation also results in replenishment of soil nitrogen removed by agricultural processes. Some bacteria fix nitrogen in symbiotic associations in plants. Other Nitrogen-fixing bacteria are free-living in soil and aquatic habitats. Waste management, whether in compost, landfills or sewage treatment facilities, exploits activities of microbes in the carbon cycle [1]. Organic materials are digested by microbial enzymes into substrates that eventually are converted to a few organic acids and carbon dioxide.

The uniqueness of microorganisms and their biosynthetic capabilities, given a specific set of environmental and cultural conditions, has made them likely candidates for solving particularly difficult problems in the life and other fields as well. The various ways in which microorganisms have been used over the past 50 years to advance medical technology, human and animal health, food processing, food safety and quality, genetic engineering, environmental protection, agricultural biotechnology, and more effective treatment of agricultural and municipal wastes. Nevertheless, while microbial technologies have been applied to various agricultural and environmental problems with considerable success in recent years, they have not been widely accepted by the scientific community because it is often difficult to consistently reproduce their beneficial effects [2]. Microorganisms are effective only when they are presented with suitable and optimum conditions for metabolizing their substrates Including available water, oxygen (depending on whether the microorganisms are obligate aerobes or facultative anaerobes), pH and temperature of their environment. Meanwhile, the various types of microbial cultures and inoculants available in the market today have rapidly increased because of these new technologies. Therefore, it is necessary that future agricultural technologies be compatible with the global ecosystem and with solutions to such problems in areas different from those of conventional agricultural technologies. An area that appears to hold the greatest promise for technological advances in crop production, crop protection, and natural resource conservation is that of beneficial and effective microorganisms applied as soil, plant and environmental inoculants [3]. Since microorganisms are useful in eliminating problems associated with the use of chemical fertilizers and pesticides, they are now widely applied in nature farming and organic agriculture [4]. Environmental pollution, caused by excessive soil erosion and the associated transport of sediment, chemical fertilizers and pesticides to surface and groundwater, and improper treatment of human and animal wastes has caused serious environmental and social problems throughout the world. Often engineers have attempted to solve these problems using established chemical and physical methods. However, they have usually found that such problems cannot be solved without using microbial methods and technologies in coordination with agricultural production [5]. For many years, soil microbiologists and microbial ecologists have tended to differentiate soil microorganisms as beneficial or harmful according to their functions and how they affect soil quality, plant growth and yield, and plant health [6]. Role of microbes are in shifting the soil microbiological equilibrium in ways that can improve soil quality, enhance crop production and protection, conserve natural resources, and ultimately create a more sustainable agriculture and environment [7, 8].

We depend on having a microbial population for our health. There are normally 10 times more microbial cells on/in a human body than human cells. The good bacteria that live on and in us protect us from the bad invaders that might come our way. Our own digestion has also evolved to use bacteria for assistance, allowing us to gain nutrition from plant polysaccharides that our own enzymes will not degrade. This is one of the reasons why taking antibiotics can lead to episodes of diarrhea - we have disrupted our natural gut biota. Microbes provide food for us through fermentation. They have been used for centuries to provide us with food. Bread is the result of a microbial fermentation of sugars to produce carbon dioxide the bubbles that make bread rise. We owe our beer and wine to similar little yeast that convert sugars into alcohol for our consumption. Yogurt and cheese are produced by bacterial fermentation of lactose, the sugar in milk. Microbes such as phytoplankton also serve as the nutrient source that indirectly feeds all marine animals. And microbial symbioses with plants allow them to grow strong and increase productivity sometimes they are even essential for plant survival [9]. Microbes can eat waste and pollution. Microorganisms are responsible for getting rid the waste generated by industry and households. They detoxify acid mine drainage and other toxins that we dump into the soil and water. The nutrients gained from the breakdown of these products then go to feed plants or algae, which in turn feed all animals. Microbes purify waste water. When we flush things down the drain or the toilet, they go to a septic system or waste water treatment plant at the end of the line. After primary mechanical treatment and aeration microbes remove organic materials from the filthy waters that flow into these systems and, eventually, water can safely be returned to the rivers and streams [10]. In this way we see the influence of microbes on the earth's environment and inhabitants.

CONCLUSION

Microbes have great impacts in the fields of health, food/ agriculture and environmental clean-up. Microbe provides an unlimited opportunity to solve problems of hunger, food security, diseases and also environmental pollution amongst the growing population in developing countries. Anaerobic bio-hydrogenation can be done by the use of microbes. Anaerobic bio-hydrogenation from biomass is receiving recent attention as one of the new sources of energy production, because wastewater and other biomass can be used as raw material and hydrogen gas can be continuously produced in an anaerobic fermenter. By the use of microbes we can generate major economic, health and environmental benefits [11]. The effects of microbes derive from their metabolic activities in the environment, their associations with plants and animals, and from their use in food production and biotechnological processes. This chapter is much interesting and useful for biotechnologists and industries because it focuses on applications of microbes and involved in management of agricultural and environmental problems.

REFERENCES

1. Parr, J.F., and Hornick, S.B. (1992). Utilization of municipal wastes. p.545-559.

2. Higa, T. (1991). Effective microorganisms: A biotechnology for mankind. p. 8-14.

3. Higa, T., and Wididana, G.N. (1991). The concept and theories of effective microorganisms. p. 118-124.

4. Higa, T., and Wididana, G.N. (1991). Changes In the soil microflora Induced by effective microorganisms. p.153-162.

5. Hornick, S.B. (1992). Factors affecting the nutritional quality of crops. Amer. J. Alternative Agric., 7; 63-68.

6. Parr, J.F., Papendick, R.I., Hornick, S.B., and Meyer, R.E. (1992). Soil quality: Attributes and relationship to alternative and sustainable agriculture. Amer. J. Alternative Agric., 7; 5-11.

7. Higa, T. (1994). Effective Microorganisms: A New Dimension for Nature Farming. p. 20-22.

8. Reganold, J.P., Papendick, R.I., and Parr, J.F. (1990). Sustainable Agriculture. Scientific American 262; (6): 112-120.

9. Parr, J.F., and Hornick, S.B. (1992). Agricultural use of organic amendments: A historical perspective. Amer. J. Alternative Agric., 7; 181-189.

10. http://benhascience.com/wordpress/?p=647. (2009). Beneficial functions of microorganisms. April 10th.

11. De M. (2005). Recent trends in biotechnology. Current Science, 88, 7; 1030-1031.

LAP LAMBERT ACADEMIC PUBLISHING AG & CO. KG, DUDWELLER LANDSTR, GERMANY

Figure 1.1: Environmental science and microbial applications.

CHAPTER-2

ECOFREINDLY APPROACH FOR NANOPARTICLE SYNTHESIS

Garima Singhal[1] and Ashish Ranjan Sharma[2]

[1]Amity Institute of Nanotechnology, Amity University Uttar Pradesh, Noida, U.P. (201301)
[2]Infectious Disease Medical Research Center, College Of Medicine, Hallym University, Chuncheon, Gangwan Do, South Korea (200702)

ABSTRACT

Till last few years synthesis of nanoparticles have been purely physical and chemical process. But recently researchers have paid attention towards biological synthesis of nanoparticles as it is environment friendly. In biological synthesis nanoparticles are synthesized using plants and microorganism. One approach that shows great potential is synthesis of nanoparticles using micro-organisms such as bacteria, yeast and fungi. Nanoparticles synthesized from biological means are also used in various applications. In this chapter exploitation of microorganisms for nanoparticle synthesis has been studied.

INDEX

INTRODUCTION

Nanotechnology is the study of compounds 1 nanometer to 100 nanometers in one dimension. Nanoparticles have unusual optical, chemical and electrical properties which make them very important. Nanoparticles are used in many respects in biotechnology and industries, for example in the fields of optics, life sciences, pharmacy, medicine, mechanics, magnetism, catalyst and energy science etc. Metal nanoparticles are very fine and strong particles which have many applications in different fields like medical imaging (Lee et al 2008), electronics (Lipovskii et al 2008), nanocomposites (Ting et al 2007), biolabeling (Tan et al 2006; Parak et al 2005), drug delivery (Horcajada et al 2008), biocide or antimicrobial agents (Sanpui et al 2008; Kirchner et al. 2004), filters (Boskovic et al 2008), non-linear optics (Ebothe et al 2006), hyperthermia of tumors (Pissuwan et al. 2006), intercalation materials for electrical batteries (Joerger et al 2001), sensors (Jiang et al 2008), optical receptors (Dahan et al 2003), catalysis in chemical reactions (Kralik et al 2000), and etc (Nam and Lead 2008; Bhattacharya and Mukherjee 2008; Parak et al 2003; Pellegrino et al 2005). Various methods are used for synthesizing metal nanoparticles. Three main nanoparticle synthesis techniques have been developed: chemical synthesis (Masala and Seshadri 2004), Physical synthesis (Swihar 2003) and biological synthesis (Konishi 2006). Some of the well known methods are chemical recovery using regenerative materials, aerosol technique, electrochemical deposition, photochemical recovery and laser exposure. Green nanotechnology has gained the attention of researchers towards ecofriendly ways for nanoparticle synthesis which includes a wide range of processes that reduce or eliminate toxic substances to restore environment. Green Nanotechnology also seek more effective alternatives for energy production (e.g. solar and fuel cells). In this approach nanoparticles are synthesized using biological materials like microorganisms, plants etc. Some biological molecules like fatty acids, amino acids, are also used as template in the growth of semiconductor nanocrystals. Particularly biological materials like DNA (Berti, 2008), protein cages (Flenniken and Uchida, 2009), biolipid cylinders, viroid capsules, S-layers and multicellular superstructures have been used in template-mediated synthesis of inorganic nanoparticles. In green nanotechnology, microorganisms can be exploited for the synthesis of nanoparticles.

Advantages of biosynthesis of nanoparticles over chemical and physical methods of synthesis include:

1. Biological nanoparticle synthesis is cost effective in terms of greater commercial viability and large savings in energy costs and high production rate in comparison with conventional methods (Mukherjee et al 2004).

2. Biological synthesis can be successfully used for production of small nanoparticles in large-scales operations while large-scale production by chemical and physical methods usually results in particles larger than several micrometers (Klaus et al 1999).

3. It is a clean, nontoxic and eco-friendly method (Senapati et al 2005).

4. Physical and chemical methods need sophisticated instrumentation to provide high temperature and high pressure which is a harder situation to provide. (Bansal et al 2004).

As we all know that microorganisms defend themselves from metal toxicity by reducing the metals and converting them in non toxic forms. Thus, this capability of microorganisms can be approached for synthesizing metal nanoparticles in a non toxic way. Among microorganisms both eukaryotes and prokaryotes are known for nanoparticle synthesis capability. Monodispersity, size and shape of the nanoparticles can be controlled by various methods to get the best particles. However, there is a growing need to understand the basics of this technique to facilitate application of the new methodology to laboratory and industrial needs.

The reaction condition and quality of nanoparticle can be optimized by changing experimental factors such as pH, incubation time, Concentration of precursor, presence of light source, temperature, the composition of the culture medium, etc. This optimization will improve the quality i.e. chemical composition, shape and size, and monodispersity of the particles synthesized (Klaus et al 1999). For example, at pH 7 gold nanoparticles of 10-20 nm were synthesized in the periplasmic space of mesophilic bacterium *Shewanella algae* cells. When the solution pH was decreased to 1, gold nanoparticles of 50-500 nm were precipitated outside of cells. Therefore, the size and location of gold nanoparticles were controlled here by optimizing pH of reaction mixture (Konishi et al 2006). In another example of biosynthesis of gold nanoparticles by *Rhodopseudomonas capsulata*, variations in morphology of nanoparticles was observed from spherical shape of 10-20 nm at pH 7, to nanoplates at pH 4 (He et al 2007). In preparation of silver nanoparticles by *Plectonema boryanum* variation of temperature controlled the range of particles size (Lengke et al 2007).

For utilization in different scientific fields, biological synthesis still needs optimized reaction conditions, and also an understanding of the biochemical and molecular mechanisms of the reaction with the help of which better chemical composition, shape, size, and monodispersity can be obtained. This chapter contains a brief outlook of the processes and classification of environment friendly approach for metal nanoparticles biosynthesis by different microorganisms.

CLASSIFICATION

Metal nanoparticle biosynthesis techniques are classified as either intracellular or extracellular biosynthesis. In extracellular biosynthesis biological reagents required for the bioreduction are presented in bioliquids (Kalishwaralal et al 2008). Here, bioreduction of metal nanoparticles does not take place in microbial growth phases. Bioliquids used for biosynthesis can be either of the following:

1. The supernatant of cultures which are prepared from centrifuging the microbial culture after its growth (Shahverdi et al 2007).

2. Sterile supernatant which comes from sterilizing the supernatant by filter which makes it completely free from microbes as if microbial growth happens, the mode changes to intracellular (Husseiny et al 2008).

3. Water containing cell biomass (Durán et al 2005).

4. Water which has kept biomass for a day (Ahmad et al 2003a).

Extracellular biosynthesis is further classified in two different preparation methods: rapid synthesis and slow synthesis. As the name suggests, the former (rapid) can be done in a few minutes, while the latter (slow) requires several hours or even days. As an example of rapid synthesis, culture supernatant of *Klebsiella pneumonia* is used for formation of silver nanoparticles in 5 minutes (Shahverdi et al 2007), while its formation in 24 hours by mycelia mat of *Phaenerochaete chrysosporiom* has been classified as a slow synthesis (Vigneshwaran et al 2006).

In intracellular biosynthesis time limiting factor is in-vivo synthesis in cells. Metals are bioreduced and deposited inside the cell by enzymatic processes. This process is also known as detoxification of hazardous materials. Nanoparticles synthesized intracellularly can be separated from the cells after getting synthesized by designed method (Ahmad et al 2003b).

Advantages of extracellular biosynthesis over intracellular biosynthesis are:

1. In extracellular biosynthesis nanoparticles are formed outside the biomass in the extracellular fashion, so there is no need for an additional step of processing to release the nanoparticles from the biomass by ultrasound treatment or by reaction with suitable detergents.

2. Extracellular biosynthesis is a cheaper and simpler downstream processing.

Biosynthesis of nanoparticles can be done either extracellularly or intracellularly. Basic steps for metal nanoparticle biosynthesis are the same. First, culture medium supplemented with metal ion should be prepared. For example for gold, $HAuCl_4$, or for silver, $AgNO_3$

must be added to culture as metal resource. Second, the prepared culture media should be inoculated with the selected microorganisms for nanoparticle bioreduction. During the different phases of microbial growth, intercellular or extracellular bioreductant ingredients perform the metal reduction process (Marcato et al 2005).

REACTION MECHANISM

Biological synthesis or biosynthesis of nanoparticles refers to the phenomena which takes place by means of biological processes or enzymatic reactions. Biosynthesis leads to "Green nanotechnology" which is an eco-friendly approach used to obtain better metal nanoparticles from microbial cells (Mandal et al 2006) and plants (Safaepour et al 2009). The biochemical mechanism of the process of biosynthesis is still unexplored and needs to be investigated. The studies of the biomaterial actually performing the reaction like enzyme structure and the genes which code these enzymes may help improve our understanding of how metal nanoparticle synthesis is performed. With these findings we can improve the chemical composition, size and shape and dispersity of generated nanoparticles which could allow the use of nanobiotechnology in a variety of other applications (Bharde et al 2007; Bharde et al 2006; Roh et al 2006).

Microorganisms have the ability to fight stress; therefore, they grow in high metal ion concentration. Applications of microbial resistance to high metal concentration includes bioleaching (Rohwerder et al 2004; Olson et al 2004) of ores, bioremediation of waters (Satinder et al 2006; Ian et al 2003) etc. Besides these application researchers have paid attention towards metal nanoparticle synthesis. Microorganisms fight stress by mechanisms including: efflux systems; bioabsorption; bioreduction; bioaccumulation; extracellular complexation or precipitation of metals; alteration of solubility and toxicity via reduction or oxidation; and lack of specific metal transport systems (Bruins et al 2000; Beveridge et al 1997). Mostly studies reaction mechanism is bioreduction in which specific reducing enzymes like NADH-dependent reductase or nitrate-dependent reductase are used by the microbial cell to reduce the metal ions (Mandal et al 2006). Protein assays indicate that main factor responsible for biosynthesis processes are enzymes of oxido-reductase group. Example of such a well known enzyme is NADH-dependent reductase. This reductase gains electrons from co-enzyme NADH and oxidizes it to NAD^+. The enzyme is further oxidized by the simultaneous reduction of metal ions into elemental form (Senapati et al 2005). Another example of such important enzyme that is responsible for this reduction in some microorganisms is nitrate-dependent reductase. In microorganism like *Fusarium oxysporum*, this enzyme conjugates with an electron donor (quinine) and reduces the metal ion into elemental form (Durán et al 2005). It is assumed that complex electron transport systems are involved in the case of rapid extracellular biosynthesis process as the reduction happens in very few minutes. In chemical method of synthesis capping or stabilizing agent is required along with a reducing agent (Raveendran et al 2005). While in biosynthesis requirement of capping agent to prevent the aggregation of fine particles is eliminated this can be clearly indicated by TEM images of nanoparticle samples synthesized. This may be because of the stabilizing capping proteins, which are secreted from microorganisms. One example of such important enzyme is Cytochrome C (Mukherjee et al 2004).

BIOLOGICAL SYNTHESIS OF NANOPARTICLE

Various microorganisms can be used for nanoparticles synthesis such as bacteria *Pseudomonas aeruginosa* (Husseiny et al 2007) and *Escherichia coli* (Du et al 2007)), fungi *Fusarium oxysporum* (Ahmad et al. 2003) and *Aspergillus fumigatus* (Bhainsa and D'Souza 2006)), algae such as *Sargassum wightii* (Singaravelu et al 2007) etc.. Fungi have certain advantages over other microorganisms used such as they are easy to handle as compared to other classes of microorganisms (Sastry et al 2003). On the other hand fungi also have a disadvantage that metal nanoparticle colloids may be contaminated with some macromolecules, such as proteins, which is present in the fungal mycelial mat (Basavaraja et al 2008).

Microorganisms may produce metal based particles in elemental form, oxide form or sulfide forms. Examples of elemental forms include gold and silver nanoparticles (Mukherjee et al 2001; Kalimuthu et al 2008). In oxide form nanoparticles of titanium and iron nanoparticles are reported (Moon et al 2007a; Moon et al 2007b; Bansal et al 2005) while zinc and cadmium nanoparticles are obtained in sulfide form (Flenniken et al 2004; Bai et al 2006).

NANOPARTICLE SYNTHESIS USING BACTERIA

Bacterial cells can be used for synthesizing nanoparticles. Both gram positive and gram negative bacterial cells are found to synthesize nanoparticles. Some of the known bacterial species for this purpose are:

1. Gram negative bacteria: Bacterium *Pseudomonas stutzeri AG259*, a silver mine bacterium, can produce silver nanocrystals in periplasmic space (Karbasian et al 2008; Mandal et al 2006; Mohanpuria et al, 2008) and also reduce selenite into insoluble elemental selenium aerobically (Garbisu et al 1996). Recently, extracellular biosynthesis of gold nanoparticles was also observed by cell supernatant of P. aeruginosa (Mohanpuria et al 2008; Husseiny et al 2008). Enterobacteriaceae is a class of enteric bacteria and includes bacterial species like *Klebsiella pneumonia*, *E. coli* and *Enterobacter cloacae*. This group was found to synthesize silver nanoparticles by reducing Ag^+ to Ag^0 (Murali 2003). *E. coli* DH5α can bioreduce chloroauric acid to Au^0 nanoparticles on the cell surface which were mostly spherical with little other morphologies of triangles and quasihexagons (Narayanan and Sakthivel 2010). CdS nanoparticles were found to accumulate intracellularly by *E. coli* and a strain of *Klebsiella pneumoniae* (Krumov et al 2009, Baia et al 2009; Holmes et al 1997). A facultative anaerobic bacterium, *Enterobacter cloacea* can also bioreduce selenite to selenium (Narayanan and Sakthivel 2010). *Geobacter ferrireducens* is a Fe (III) reducing bacterium which has the ability to precipitate gold intracellularly in periplasmic space (Narayanan and Sakthivel 2010). *Thermoanaerobacter ethanolicus* (TOR-39), a thermophilic iron reducing bacteria, synthesized magnetic octahedral nanoparticles of sizes

LAP LAMBERT ACADEMIC PUBLISHING AG & CO. KG, DUDWELLER LANDSTR, GERMANY

below 12 nm extracellularly on the surface of the strain (Krumov et al 2009; Moon et al 2007b). Magnetotactic bacteria such as *A. magnetotacticum*, Magnetotactic bacterium *MV-1*, Sulfate reducing *bacteria*, *M. magnetotacticum* and *M. gryphiswaldense* are also known for synthesizing nanoparticles. *Magnetospirillum magneticum* can produce two types of particles; magnetite (Fe_3O_4) and greigite (Fe_3S_4) nanoparticles (Mohanpuria et al 2008). Crystals of ordered single-domain magnetite (Fe_3O_4) particles were produced by *Aquaspirillum magnetotacticum* Magnetite nanoparticles formed by these bacteria showed predominant morphologies of octahedral prism, parallelepiped, cubo-octahedral and hexagonal prism in the size range of 2–120 nm (Narayanan and Sakthivel 2010).

A cell free extract of *Rhodopseudomonas capsulata* has been found to synthesize Au nano wires with a network structure and the process can be controlled to change the shape of NPs by changing the concentrations. Spherical Au NPs with 10-20 nm range have been observed at lower concentrations and nano wires with a network structure at higher concentrations (He et al 2007)

2. Gram positive bacteria: *Bacillus subtilis* 168 is able to reduce Au^{+3} ions to produce octahedral gold particles of nanoscale dimensions (5–25 nm) (Southam & Beveridge 1994; Fortin & Beveridge 2000). *B. licheniformis* was also found to reduce Ag^+ ions to Ag^0 of 5–15 nm in size in the periplasmic space of the cell (Narayanan and Sakthivel 2010). *Clostridium thermoaceticum* precipitated bright yellow CdS crystals on the surfaces of the cells and also in culture mediun within 24–48 h after the addition of $CdCl_2$ and 0.05% cysteine hydrochloride (Narayanan and Sakthivel 2010; Mandal et al 2006; Bai et al 2006). *S. aureus* can bioreduce $AgNO_3$ into Ag nanoparticles, the reaction started within a few minutes and the color of the solution turned to yellowish brown, indicating the formation of AgNPs extracellularly (Nanda and Saravanan 2009). *Lactobacillus* sp. was found to synthesize stable gold, silver, and gold-silver alloy crystals of well-defined morphology (Mohanpuria et al, 2008, Sintubin and Windt, 2009). Spherical TiO_2 nanoparticles were also found to be synthesized (Jha and Prasad, 2010) which were lighter in weight and have

high resistance to corrosion (Narayanan and Sakthivel, 2010). *Desulfovibrio desulfuricans* NCIMB 8307 has been found to synthesize nanoparticles with different morphologies like palladium nanoparticles on the surface of cells and selenium nanoparticles both inside and outside the cell in the presence of exogenous electron donor. Gold (I)-thiosulfate complex was reduced to elemental gold by heterotrophic sulfate-reducing bacterial (SRB) enrichment from a gold mine (Mohanpuria et al, 2008). Also these species can form ZnS particles with a diameter of 2–5 nm (Mandal et al 2006; Kaushik et al 2010). *Desulfosporosinus* sp., a Gram-positive sulfate-reducing microbe isolated from sediments reduces hexavalent uranium U (VI) tetravalent uranium U (IV) which precipitated uraninite in the size range of 1.5–2.5 nm (Narayanan and Sakthivel 2010).

NANOPARTICLE SYNTHESIS USING YEAST

Yeasts are simple eukaryotes and are explored mostly in the biosynthesis of the nanoparticles. *Candida glabrata* and *Schizosaccharomyces pombe* converts Cd^{+2} ions intracellularly into CdS quantum dots (Mandal et al 2006; Baia et al 2009; Murali 2003). Peptide-coated Cd NPs are biosynthesized by yeast and also known as 'Semiconductor Crystals' or 'Quantum Semiconductor Crystals' studied in solid-state physics (Dameron et al., 1989). These CdS nanoparticles have advantageous features like Crystallinity of the particles, narrow size distribution, and good water solubility. Use of yeast for nanoparticles biosynthesis has an advantage that agglomeration of nanoparticles into larger particles is prevented. Stabilization and particle size control of nanocrystals derived from yeast is done by the phytochelatin layer (Krumov et al 2009). Recently yeast strain *MKY3* has been found to form silver nanoparticles of size range of 2–5 nm extracellularly (Kowshik 2003; Krumov et al 2009). Similarly *Torulopsis sp.* Converts Pb^{+2} into PbS nanocrystals intracellularly. Crystallites, which are extracted from the biomass by freeze thawing, are 2–5 nm in size (Mandal et al 2006). Gold nanoparticles are also known to be formed by yeast strains by controlling growth and other cellular activities controlled size and shape of the nanoparticles was achieved (Mohanpuria et al 2008).

LAP LAMBERT ACADEMIC PUBLISHING AG & CO. KG, DUDWELLER LANDSTR, GERMANY

NANOPARTICLE SYNTHESIS USING FUNGI

Fungi are also explored for the biosynthesis of nanoparticles and their subsequent applications. Fungi have a wide range of diversity and have advantages over other organisms that they are easy to isolate and culture, and secrete large amounts of extracellular enzymes. Therefore, downstream processing and handling of the biomass is much simpler then bacteria as it is easily filtered by filter press of similar simple equipment (Murali 2003). Fungi are known for its ability of bioremediation as they hydrolyze metal ions quickly by many of the proteins secreted by them in a non-hazardous processes way. In addition, nanoparticles obtained from fungi are of high monodispersity and dimensions

(Rai and Yadav 2009; Mukherje and Ahmad 2001). Example of such fungi is *Verticillium sp.* When it is exposed to aqueous $AgNO_3$ it causes the reduction of the metal ions and formation of silver nanoparticles of about 25 nm diameters (Karbasian et al 2008). In the same way spherical and rod shaped gold nanoparticles are obtained by *Verticillium luteoalbum* (Narayanan and Sakthivel 2010) and by *F. oxysporum* extracellularly (Mohanpuria et al 2008). *F. oxysporum* also biosynthesized silver nanoparticles (Karbasian et al 2008; Saifuddin et al 2009; Durán et al 2005). Bimetallic Au–Ag alloy, silica, titania, zirconia, magnetite, Bi_2O_3, platinum nanoparticles, Cadmium Carbonate, Cadmium sulfide, Strontium Carbonate etc can also be synthesized using fungi (Mohanpuria et al 2008; Rautaray et al 2004; Flenniken and Uchida 2009; Lengke et al 2006; Durán et al 2005; Ahmad et al 2002; Sanyal et al 2005; Mandal et al 2006; Rai and Yadav 2009). Other fungi reported for Silver nanoparticles synthesis are *Trichoderma asperellum* and *Penicillium* sp., *Aspergillus niger* (Rai and Yadav 2009), *Aspergillus flavus* (Mohanpuria et al 2008, Sintubin & Windt 2009) and *Aspergillus fumigates* (Bhainsa and D'Souza 2006). Filamentous fungus *Hormoconis resinae* also can synthesis silver nanoparticles (Varsheny et al 2009).

LAP LAMBERT ACADEMIC PUBLISHING AG & CO. KG, DUDWELLER LANDSTR, GERMANY

NANOPARTICLE SYNTHESIS USING ACTINOMYCETES

Actinomycetes share important characteristics of fungi and bacteria (Murali 2003). Gold ions are reduced by an actinomycetes *Thermomonospora* sp. extracellularly, yielding gold nanoparticles with a much improved polydispersity (Mohanpuria et al 2008,Ahmad et al 2003a). *Rhodococcus* sp., an alkalophile, produced intracellular gold nanoparticles of the dimension 5-15 nm with good monodispersity aggregated on the plasma membrane (Krumov et al 2009).

NANOPARTICLE SYNTHESIS USING ALGAE

Biosynthesis of nanoparticles using algae has been also studied. It was observed that gold nanoparticles have been synthesized extracellularly by algae *Kappaphycus alvarezii* (Rajasulochana et al 2010) and *Sargassum wightii* (**Singaravelu et al 2007**). Gold nanoparticles were synthesized rapidly during stationary phase and were spherical in shape and morphology (Rajasulochana et al 2010). Moreover the nanoparticles were not toxic to the cells as the cells kept on growing even after nanoparticle synthesis. With the help of TEM images it was seen that well monodispersed nanoparticles of 8-12 nm size were formed (**Singaravelu et al 2007**).

NANOPARTICLE SYNTHESIS USING VIRUSES

Viruses have been explored as template for nanoparticles synthesis like tobacco mosaic virus (TMV) was used for the synthesis of iron oxides nanoparticle by oxidative hydrolysis. TMV has also been used as template for co-crystallization of CdS and PbS, and the synthesis of SiO_2 by sol–gel condensation. Protein coat present in viruses contains external groups of glutamate and aspartate on the external surface which facilitates nanoparticle formation (Narayanan and Sakthivel 2010; Pugazhenthiran et al 2009; Varsheny et al 2009; Ahmad and Senapati 2003). For nucleation and assembly of inorganic materials viral scaffold serves as template. Other viruses explored for nanoparticles formation are, cowpea chorotic mottle virus and cowpea mosaic virus (Parikh et al 2008).

LAP LAMBERT ACADEMIC PUBLISHING AG & CO. KG, DUDWELLER LANDSTR, GERMANY

REFRENCES

1. Ahmad, A., Mukherjee, P. & Mandal, D. (2002). Enzyme mediated extracellular synthesis of CdS nanoparticles by the fungus, *Fusarium oxysporum*. *Journal of American Chemical Society, 124(41)*, 12108-9.

2. Ahmad, A., Senapati, S., Khan, M.I., Kumar, R., Ramani, R., Srinivas, V. & Sastry, M. (2003b). Intracellular synthesis of gold nanoparticles by a novel alkalotolerant actinomycetes *Rhodococcus species*. *Nanotechnology, 14*, 824-828.

3. Ahmad, A., Mukherjee, P., Mandal, D., Senapati, S., Khan, M.I., Kumar, R. & Sastry, M. (2003a). Extracellular biosynthesis of silver nanoparticles using the fungus *Fusarium oxysporum* composite metal particles, and the atom to metal. *Colloids and Surfaces B: Biointerfaces, 28*, 313-318.

4. Bai, H.J., Zhang, Z.M. & Gong, J. (2006). Biosynthesis synthesis of semiconductor zinc sulfide nanoparticles by immobilized *Rhodobacter sphaeroides*. *Biotechnol lett, 28*, 1135-1139.

5. Baia, H.J., Zhangb, Z.M. & Guo, Y. (2009). Biosynthesis of cadmium sulfide nanoparticles by photosynthetic bacteria *Rhodopseudomonas palustris*. *Colloids and Surfaces B: Biointerfaces, 70*, 142–146.

6. Bansal, V., Rautaray, D., Ahmad, A. & Sastry, M. (2004). Biosynthesis of zirconia nanoparticles using the fungus *Fusarium oxysporum*. *Journal of Materials Chemistry, 14*, 3303-3305.

7. Bansal, V., Rautaray, D., Bharde, A., Ahire, K., Sanyal, A., Ahmad, A. & Sastry, M. (2005). Fungus mediated biosynthesis of silica and titania particles. *Journal of Materials Chemistry, 15*, 2583-2589.

8. Basavaraja, S.S., Balaji, S.D., Lagashetty, A.K., Rajasab, A.H. & Venkataraman, A. (2008). Extracellular biosynthesis of silver nanoparticles using the fungus *Fusarium semitectum*. *Materials Research Bulletin, 43*,1164-1170.

9. Berti, L. (2008). Nucleic acid and nucleotide-mediated synthesis of inorganic nanoparticles. *NatNanotechnol, 3(2)*, 81-7.

10. Beveridge, T.J., Hughes, M.N., Lee, H., Leung, K.T., Poole, R.K., Savvaidis, I., Silver, S. & Trevors, J.T. (1997). Metal–microbe interactions: contemporary approaches. *Advances in Microbial Physiology, 38,* 177–243.

11. Bhainsa, K.C. & D'Souza, S.F. (2006). Extracellular biosynthesis of silver nanoparticles using the fungus *Aspergillus fumigatus. Colloids and Surfaces B: Biointerfaces, 47,* 160–164.

12. Bharde, A., Rautaray, D., Bansal, V., Ahmad, A., Sarkar, I. & Yusuf, S.M. (2006). Extracellular Biosynthesis of Magnetite using Fungi. *Small, 2(1),* 135 – 141.

13. Bharde, A., Kulkarni, A., Rao, M., Prabhune, A. & Sastry, M. (2007). Bacterial Enzyme Mediated Biosynthesis of Gold Nanoparticles. *Journal of Nanoscience and Nanotechnology, 7,* 1–9.

14. Bhattacharya, R. & Mukherjee, P. (2008). Biological properties of naked metal nanoparticles. *Advanced Drug Delivery Reviews, 60(11),* 1289-1306.

15. Boskovic, L., Agranovski, I.E., Altman, I.S., & Braddock, R.D. (2008) Filter efficiency as a function of nanoparticle velocity and shape. *Journal of Aerosol Science, 39(7),* 635-644.

16. Bruins, R.M., Kapil, S. & Oehme, S.W. (2000). Microbial resistance to metal in the environment. *Ecotoxicology and Environmental Safety, 45,* 98–207.

17. Dahan, M., Lévi, S., Luccardini, C., Rostaing, P., Riveau, B., & Triller, A. (2003). Diffusion Dynamics of Glycine Receptors Revealed by Single-Quantum Dot Trackin. *Science, 302(5644),* 442 – 445.

18. Dameron, C.T., Reese, R.N., Mehra, R.K., Kortan, A.R., Carroll, P.J., Steigerwald, M.L., Brus, L.E. & Winge, D.R. (1989). Biosynthesis of cadmium sulfide quantum semiconductor crystallites. *Nature, 338,* 596–597

19. Du, L., Jiang, H., Liu, X. & Wang, E. (2007). Biosynthesis of gold nanoparticles assisted by Escherichia coli DH5a and its application on direct electrochemistry of hemoglobin. *Electrochemistry Communications, 9,* 1165-1170.

LAP LAMBERT ACADEMIC PUBLISHING AG & CO. KG, DUDWELLER LANDSTR, GERMANY

20. Duran, N., Marcato, P.D., Alves, O.L., Souza, I. & Esposito, E. (2005). Mechanistic aspects of biosynthesis of silver nanoparticles by several *Fusarium oxysporum* strains. *Journal of Nanobiotechnology, 3,* 8-14.

21. Ebothe, J., Kityk, I.V., Chang, G., Oyama, M. & Plucinski, K.J. (2006). Pd nanoparticles as new materials for acoustically induced non-linear optics. *Physica E: Low-dimensional Systems and Nanostructures, 35(1),* 121-125.

22. Flenniken, M.L. & Uchida, M. (2009). A Library of Protein Cage Architectures as Nanomaterials, Viruses and Nanotechnology. *Current Topics in Microbiology and Immunology, 327,* 71-93.

23. Flenniken, M.M., Allen, T. & Douglas, (2004). Microbe Manufacturers of Semiconductors. *Chemistry & Biology, 11(11),* 1478 – 1480.

24. Fortin, D. & Beveridge, T.J. (2000). From biology to biotechnology and medical applications. *In: Aeuerien E (ed) Biomineralization, Wiley-VCH, Weinheim,* 7- 22.

25. Garbisu, C., Ishiia, T., Leightonb, T. & Buchanana, B.B. (1996). Bacterial reduction of selenite to elemental selenium. *Chemical Geology, 132(1-4),* 199-204.

26. He, S., Guo Z. & Zhang, Y. (2007). Biosynthesis of gold nanoparticles using the bacteria *Rhodopseudomonas capsulata. Materials Letters, 61,* 3984–3987.

27. Holmes, J.D., Richardson, D.J. & Saed, S. (1997) Cadmium-specific formation of metal sulfide 'Q-particles' by *Klebsiella pneumoniae. Microbiology, 143,* 2521-30.

28. Horcajada, P., Serre, C., Maurin, G., Ramsahye, N.A., Balas, F., Vallet-Regí, M. & Sebban, M. (2008). Flexible Porous Metal-Organic Frameworks for a Controlled Drug Delivery. *Journal-American Chemical Society, 130(21),* 6774–6780.

29. Husseiny, M.I., Ei-Aziz, M.A., Badr, Y. & Mahmoud, M.A. (2008). Biosynthesis of gold nanoparticles using *Pseudomonas aeruginosa. Spectrochimica Acta Part A: Molecular and Biomolecular Spectroscopy, 67,* 1003- 1006

30. Ian, H., Singleton, I. & Milner, M.G. (2003). Bioremediation: A Critical Review. *Newcastle: Horizon Scientific Press.*

31. Jha, A.K. & Prasad, D. (2010). Biosynthesis of metal and oxide nanoparticles using *Lactobacilli* from yoghurt and probiotic spore tablets. *Biotechnology, 5(3)*, Pages 285 – 291.

32. Jiang, K., Liu, W., Wan, L. & Zhang, J. (2008). Manipulation of ZnO nanostructures using dielectrophoretic effect. *Sensors and Actuators B: Chemical, 134(1)*, 79-88.

33. Joerger, T.K., Joerger, R., Olsson, E. & Granqvist, C.G. (2001). Bacteria as workers in the living factory: metal-accumulating bacteria and their potential for materials science. *Trends in Biotechnology, 19(1)*, 15-20.

34. Kalimuthu, K., Sureshbabu, R., Venkatraman, D., Bilal, M. & Gurunathan, S. (2008). Biosynthesis of silver nanocrystals by *Bacillus licheniformis*. *Colloids and Surfaces B: Biointerfaces, 65(1)*, 150-153.

35. Kalishwaralal, K., Deepak, V., Ramkumarpandian, S., Nellaiah, H. & Sangiliyandi, (2008). Extracellular biosynthesis of silver nanoparticles by the culture supernatant of *Bacillus licheniformis*. *Materials Letters, 62*, 4411-4413.

36. Karbasian, M., Atyabi, S.M. & Siadat, S.D. (2008) Optimizing Nano-silver Formation by *Fusarium oxysporum*. *American Journal of Agricultural and Biological Science, 3(1)*, 433-437, 2008.

37. Kaushik, N., Thakkar, S., Mhatre, S. & Parikh, R.Y. (2010). Biological synthesis of metallic nanoparticles. *Nanotechnology, Biology, and Medicine, 6*, 257.

38. Kirchner, C., Liedl, T., Kudera, S., Pellegrino, T., Muñoz Javier, A., Gaub, H.E., Stölzle, S., Fertig, N. & Parak, W.J. (2004). Cytotoxicity of Colloidal CdSe and CdSe/ZnS Nanoparticles. *Nano Letters, 5(2)*, 331-338.

39. Klaus, T., Joerger, R., Olsson, E. & Granqvist, C.G. (1999). Silver-based crystalline nanoparticles, microbially fabricated. *The Proceedings of the National Academy of Sciences Online (US), 96(24)*, 13611–13614.

LAP LAMBERT ACADEMIC PUBLISHING AG & CO. KG, DUDWELLER LANDSTR, GERMANY

40. Konishi, Y., Ohno, K., Saitoh, N., Nomura, T. & Nagamine, S. (2006). Microbial Synthesis of Noble Metal Nanoparticles Using Metal-Reducing Bacteria. *Journal of the Society of Powder Technology, 43(7),* 515-521.

41. Kowshik, M. (2003). Ashataputre.Extracellular synthesis of silver nanoparticles by a silver-tolerant yeast strain MKY3. *Nanotechnology, 14,* 95–100.

42. Kralik, M., Corain, B. & Zecca, M. (2000). Catalysis by Metal Nanoparticles Supported on Functionalized Polymers. *chemistry papers, 54(6b),* 254-264.

43. Krumov, N., Perner-Nochta, I., Oder, S. & Gotcheva, V. (2009). Production of Inorganic Nanoparticles by Microorganisms. *Chem. Eng. Technol., 32(7),* 1026–1035.

44. Lee, H.Y., Li, Z., Chen, K., Hsu, A.R., Xu, C.J., Xie, J., Sun, S.H. & Chen, X.Y. (2008). PET/MRI dual-modality tumor imaging using arginine-glycine-aspartic (RGD)-conjugated radiolabeled iron oxide nanoparticles. *Journal of Nuclear Medicine, 49(8),* 1371-1379.

45. Lengke, M.F., Fleet, M. E. & Southam, G. (2007). Biosynthesis of Silver Nanoparticles by Filamentous Cyanobacteria from a Silver(I) Nitrate Complex. *Langmuir, 23,* 2694-2699.

46. Lengke, M.F., Fleet, M.E. & Southam, G. (2006). Synthesis of Platinum Nanoparticles by Reaction of Filamentous Cyanobacteria with Platinum(IV)–Chloride Complex. *Langmuir, 22 (17),* 7318–7323.

47. Lipovskii, A.A., Kuittinen, M., Karvinen, P., Leinonen, K., Melehin, V.G., Zhurikhina, V.V. & Svirko, Y.P. (2008). Electric field imprinting of sub-micron patterns in glass–metal nanocomposites. *Nanotechnology, 19,* 415304-415309.

48. Mandal, D., Bolander, M.E., Mukhopadhyay, D., Sarkar, G. & Mukherjee, P. (2006). The use of microorganisms for the formation of metal nanoparticles and their application. *Applied Microbiology and Biotechnology, 69,* 485-492.

49. Marcato, P.D., De Souza, G.I.H., Alves, O.L. & Esposito, E. (2005). Antibacterial Activity of Silver Nanoparticles Synthesized by Fusarium oxysporum Strain. 2nd

Mercosur Congress on Chemical Engineering and 4th Mercosur Congress on Process Systems Engineering; Aug 14-18; Village Rio das Pedras, Club Med, Rio de Janeiro.

50. Masala, O. & Seshadri, R. (2004). Synthesis Routes for Large Volumes of Nanoparticles. *Annual Review of Materials Research, 34,* 41-81.

51. Mohanpuria, P., Nisha, K., Rana, N.K. & Yadav, S.K. (2008). Biosynthesis of nanoparticles: technological concepts and future applications. *Journal of Nanoparticle Research, 10,* 507–517.

52. Moon, J.W., Roh, Y., Lauf, R.J., Vali, H., Yeary, L.W. & Phelps, T.J. (2007a). Microbial preparation of metal-substituted magnetite nanoparticles. *Journal of Microbiological Methods, 70,* 150–158.

53. Moon, J.M., Roh, Y., Yeary, L.W. (2007b). Microbial formation of lanthanide-substituted magnetites by *Thermoanaerobacter sp.* TOR-39. *Extremophiles, 11,* 859–867.

54. Mukherjee, P., Ahmad, A., Mandal, D., Senapati, S., Sainkar, S.R., Khan, M.I., Ramani, R., Parischa, R., Ajayakumar, P.V., Alam, M., Sastry, M. & Kumar, R. (2001). Bioreduction of AuCl4- ions by the fungus, *Verticillium sp.* and surface trapping of gold nanoparticles formed. *Angewandte Chemie International Edition, 40,* 3585-3588.

55. Mukherjee, P., Senapati, S. & Mandal, D. (2004). Extracellular Synthesis of Gold Nanoparticles by the Fungus *Fusarium oxysporum. ChemBioChem, 3(5),* 461 – 463.

56. Mukherjee, P. & Ahmad, A. (2001). Fungus-Mediated Synthesis of Silver Nanoparticles and Their Immobilization in the Mycelial Matrix: A Novel Biological Approach to Nanoparticle Synthesis. *Nano Letters, 1(10),* 515–519.

57. Murali, S. (2003). Biosynthesis of metal nanoparticles using fungi and Actinomycete. *Current Science, 85,* 2.

58. Nam, Y.J. & Lead, J.R. (2008). Manufactured nanoparticles: An overview of their chemistry, interactions and potential environmental implications. *Science of the Total Environment, 400(1-3)*, 396-414.

59. Nanda, A. & Saravanan, M. (2009). Biosynthesis of silver nanoparticles from *Staphylococcus aureus* and its antimicrobial activity against MRSA and MRSE. *Nanomedicine: Nanotechnology, Biology, and Medicine, 5,* 452–456.

60. Narayanan, K.B. & Sakthivel, N. (2010). Biological synthesis of metal nanoparticles by microbes. *Advances in Colloid and Interface Science, 156,* 1–13.

61. Olson, G.J., Brierley, J.A. & Brierley, C.L. (2004). Bioleaching review part B: Progress in bioleaching: applications of microbial processes by the minerals industries. *Applied Microbiology and Biotechnology, 63(3),* 249-257.

62. Panigrahi, S., Kundu, S., Ghosh, S., Nath, S. & Pal, T. (2004). General method of synthesis for metal nanoparticles. *Journal of Nanoparticle Research, 6(4),* 411-414.

63. Parak, W.J., Gerion, D., Pellegrino, T., Zanchet, D., Micheel, C., Williams, S.C., Boudreau, R., Le Gros, M.A., Larabell, C.A. & Alivisatos, A.P. (2003). Biological applications of colloidal nanocrystals. *Nanotechnology, 14,* R15–R27.

64. Parak, W.J., Pellegrino, T. & Plank, C. (2005). Labelling of cells with quantum dots. *Nanotechnology, 16,* R9–R25.

65. Parikh, R.P., Singh, S., Prasad, B.L.V., Patole, M.S., Sastry, M. & Shouche, Y.S. (2008). Extracellular synthesis of crystalline silver nanoparticles and molecular evidence of silver resistance from *Morganella sp.*. *Chembiochem, 9(9),* 1415-22.

66. Pellegrino, T., Kudera, S., Liedl, T., Javier A.M., Manna, L. & Parak, W.J. (2005). On the Development of Colloidal Nanoparticles towards Multifunctional 6 Journal of Young Investigators January 2010 Structures and their Possible Use for Biological Applications. *Small, 1(1),* 48-63.

67. Pissuwan, D., Valenzuela, S.M. & Cortie, M.B. (2006). Therapeutic possibilities of plasmonically heated gold nanoparticles. *Trends in Biotechnology, 24(1),* 62-67.

68. Pugazhenthiran, N., Anandan, S., Kathiravan, G., Udaya Prakash, N. K., Crawford, S. & Ashokkumar, M. (2009). Microbial synthesis of silver nanoparticles by *Bacillus*. *J Nanopart, 11,*1811–1815

69. Rai, M. & Yadav, A. (2009). Myconanotechnology: a New and Emerging Science. *CAB International. Applied Mycology.*

70. Rajasulochana, P., Dhamotharan, R., Murugakoothan, P., Murugesan, S. & Krishnamoorthi, P. (2010). Biosynthesis and characterization of gold nanoparticles using the algae *Kappaphycus alvarezii. International journal of nanoscience, 9(5),* 511-516.

71. Rautaray, D., Sanyal, A. & Adyanthaya, S.D. (2004). Biological Synthesis of Strontium Carbonate Crystals Using the Fungus *Fusarium oxysporum. Langmuir, 20(16),* 6827-683.

72. Raveendran, P., Fu. J. & Wallen, S.L. (2005). A simple and green method for the synthesis of Au, Ag, and Au–Ag alloy nanoparticles. *Green Chemistry, 8,* 34–38.

73. Roh, Y., Vali, H., Phelps, T.J. & Moon, J.W. (2006). Extracellular Synthesis of Magnetite and Metal-Substituted Magnetite Nanoparticles. *Journal of Nanoscience and Nanotechnology, 6,* 3517–3520.

74. Rohwerder, T., Gehrke, T., Kinzler, K. & Sand, W. (2004). Bioleaching review part A: Progress in bioleaching: fundamentals and mechanisms of bacterial metal sulfide oxidation. *Applied Microbiology and Biotechnology, 63(3),* 239-248.

75. Safaepour, M., Shahverdi, A.R., Shahverdi, H.R., Khorramizadeh, M.R. & Gohari, A.R. (2009). Green Synthesis of Small Silver Nanoparticles Using Geraniol and Its Cytotoxicity against Fibrosarcoma-Wehi 164. *Avicenna J Med Biotech, 1(2),* 111-115.

76. Saufuddin, N., Wong, C.W. & and Nur, A.A. (2009). Rapid Biosynthesis of Silver Nanoparticles Using Culture Supernatant of Bacteria with Microwave Irradiation. *http://www.e-journals.net, 6(1),* 61-70.

LAP LAMBERT ACADEMIC PUBLISHING AG & CO. KG, DUDWELLER LANDSTR, GERMANY

77. Sanpui, P., Murugadoss, A., Durga Prasad, P.V., Ghosh, S.S. & Chattopadhyay, A. (2008). The antibacterial properties of a novel chitosan–Ag-nanoparticle composite. *International Journal of Food Microbiology, 124(2)*, 142-146.

78. Sanyal, A., Rautaray, D. & Bansal, V. (2005). Heavy-Metal Remediation by a Fungus as a Means of Production of Lead and Cadmium Carbonate Crystals. *Langmuir, 21(16)*, 7220-7224.

79. Sastry, M., Ahmad, A., Khan, M.I. & Kumar, R. (2003). Biosynthesis of metal nanoparticles using fungi and actinomycetes. *Current Science, 85,*162-170.

80. Satinder, K.M., Verma, R.Y., Surampalli, K., Misra, R.D., Tyagi, N. & Meunier, J.F.B. (2006). Bioremediation of Hazardous Wastes—A Review. *Practice Periodical of Hazardous, Toxic, and Radioactive Waste Management, 10(2)*, 59-72.

81. Senapati, S., Ahmad, A., Khan, M.I., Sastry, M. & Kumar, R. (2005). Extracellular biosynthesis of bimetallic Au-Ag alloy nanoparticles. *Small, 1*, 517-520.

82. Shahverdi, A.R., Minaeian, S., Shahverdi, H.R., Jamalifar, H. & Nohi, A.A. (2007). Rapid synthesis of silver nanoparticles using culture supernatants of Enterobacteriaceae: A novel biological approach. *Process Biochemistry, 42*, 919-923.

83. Singaravelu, G., Arockiamary, J.S., Kumar, V.G. & Govindaraju, K. (2007). A novel extracellular synthesis of monodisperse gold nanoparticles using marine alga, *Sargassum wightii Greville. Colloids and Surfaces B: Biointerfaces, 57*, 97–101.

84. Sintubin, L. & Windt, W.D. (2009). Lactic acid bacteria as reducing and capping agent for the fast and efficient production of silver nanoparticles. *Appl Microbiol Biotechnol,84*, 41–749.

85. Southam, G. & Beveridge, T.J. (1994). The in vitro formation of placer gold by bacteria *Geochim. Cosmochim. Acta, 58*, 4527–4530

86. Swihar, M.T. (2003). Vapor-phase synthesis of nanoparticles. *Current Opinion in Colloid & Interface Science, 8(1)*, 127-133.

87. Tan, M., **Wang, G., Ye, Z. & Yuan, J.** (2006). Synthesis and characterization of titania-based monodisperse fluorescent europium nanoparticles for biolabeling. *Journal of Luminescence, 117(1)*, 20-28.

88. Ting, P.K., **Hussain, Z. & Cheong, K.Y.** (2007). Synthesis and characterization of silica–titania nanocomposite via a combination of sol–gel and mechanochemical process. *Journal of Alloys and Compounds, 446(1-2)*, 304-307.

89. Varsheny, R., Mishra, A.N., Bhadauria, S. (2009). A nivel microbial rout to synthesize silver nanoparticles using fungus *Hormoconis Resinae*. *Journal of Nanomaterials and Biostructures, 4(2)*, 349 – 355.

90. Vigneshwaran, N., Kathe, A.A., Varadarajan, P.V., Nachane, R.P. & Balasubramanya, R.H. (2006). Biomimetics of silver nanoparticles by white rot fungus, *Phaenerochaete chrysosporium*. *Colloids and Surfaces B: Biointerfaces, 53*, 55-59.

LAP LAMBERT ACADEMIC PUBLISHING AG & CO. KG, DUDWELLER LANDSTR, GERMANY

CHAPTER-3

Detoxification characteristic of Probiotics- preventive role in Human cancer

Ashish Dhyani*, Tejpal Dehwa[1], Vijendra Mishra[2], Vivek Bajpai[3]

*Independent author (Molecular Microbiologist, Department of Molecular Genetics,
Nicholas Piramal India Ltd-Ex Scientist)
[1]Department of Microbiology, Dolphin (PG) Institute of Biomedical and Natural Sciences,
Manduwala, Dehradun, Uttrakhand, India-248007,
[2] Department of Dairy Microbiology, SMC College of Dairy Science, Anand University,
Gujarat, India- 388110
[3]Department of Science, FASC, MITS University, Lakshmangarh, Sikar-332311,
Rajasthan,   INDIA

Abstract

Probiotics are microbial cultures containing potentially bacteria or yeast. Since many years, there was huge research going on the beneficial effect of probiotics associated with anticarcinogenic effects. One mechanism may be detoxification of genotoxins in the gut. In great number and diversity, microbes inhabit the intestinal tract, skin, urogenital tract, nasal and oral cavity. Prebiotics are indigestible food components and that could promote the growth of beneficial bacteria including probiotics. Present studies have suggested that probiotics also possess protective effect against colon carcinogenesis. Beside the potential role in colon cancer prevention, probiotics also plays significant role in preclusion of other diseases. In this chapter we will study the various therapeutics and beneficial role of probiotics in human.

Key Words: Lactic acid bacteria, Bifidobacteria, lactose malabsorption, colon cancer, detoxification, pro and prebiotics.

INDEX

Introduction:

During the beginning of 20th century, Russian nobel laureate Dr Elie Metchnikoff observed extended longevity in Bulgarian people who were consuming large amount of fermented milk products. Since then, these consumed fermented milk products showing its remarkable benefits in human health. The human colon can be described as a complex microbial ecosystem, comprising several hundred bacterial species. Some of these enteric bacteria are beneficial to the host and have been shown to exert antimutagenic and anticarcinogenic properties[1-2].

Probiotics can be defined as living microorganisms, who confer a health benefit on the host when consumed in adequate amount[3]. Certain strains of bacteria have been discovered over the years to have probiotics properties ranging from Lactobacillus sps, streptococcus sps, Enterococcus sps, Bifidobacterium sps, etc[4]. Probiotics are also found to exert other health advantages such as improving lactose intolerance, increase humoral immune responses, biotransformation of isoflavone phytoesterogen to improve post-menopausal symptoms, bioconversion of bioactive peptides for anti-hypertension and reducing serum cholesterol level[5]. Studies from germ free animals have proven that animals do not require microbial colonization for survival, but germ free animals, compared with their conventional counter parts, demonstrate many physiologic and biochemical differences and are more susceptible to infection[6]. Thus, the difference between conventional and germ free animals have provided a basis for the belief that microbial colonization has important health implications for humans. Research on probiotics consists of experiments done with dozens of different bacterial strains and combinations of strains used at different doses in vitro, animal or human studies with dozens of different research end points. The positive results from human volunteer or clinical studies, even in the absence of compelling mechanistic studies, provide validity to the probiotic concept.

Many probiotics are present in natural sources such as lactobacillus in yogurt and sauerkraut. Claims are made that probiotics strengthen the immune system[8]. Lactic acid bacteria consist of heterogeneous group of gram-positive bacteria, whose main fermentation product from carbohydrate is lactate. The group comprises cocci (streptococcus, pediococcus, leuconostoc) and rods (lactobacillus and bifidobacterium), which are either exclusively (homofermentative) or at least 50% (heterofer-mentative) lactate producers[9].

Preventive roles on colon cancer

Probiotics plays a very significant role in prevention from many diseases in human system (Fig1). Bogdanov et al (1977) were first proved the potential anti-tumerogenic activity of L. bulgaricus. Same year, a series of epidemiological studies suggested the consumption of high levels of cultured milk products may reduce the risk of colon cancer. Colorectal cancer remains one of the leading causes of cancer morbidity and mortality among men and women in world especially in western countries. Despite several advances in the treatment of cancer, the therapeutic outcome of this malignancy is posing significant challenges to modern medicine. Consequently, primary prevention, early detection and secondary prevention are emerging as the most promising approaches for reducing the morbidity and mortality from colon cancer[18]. Early studies had indicated that the general population of intestinal bacteria is associated with initiation of colon cancer via the production of carcinogens, co-carcinogens, or pro-carcinogens[19]. Reddy et al. found that a stimulated growth of bifidobacteria in the colon could lead to the inhibition of colon carcinogenesis[20].

In general, the complex gut microflora, which consists of(>1 x10[11]) living bacteria/g colon content, Lactic acid bacteria (LAB) belongs to the beneficial group of microflora which plays a pivotal role in suppressing colon carcinogenesis by possibly influencing metabolic, immunologic, and protective functions in the colon. Population of lactic acid bacteria may increase after the consumption of foods containing probiotics; however, probiotic ingestion also increases the number and metabolic activity of LAB in the colon of humans and animals[21-26]. In animals, LAB ingestion was shown to prevent carcinogen-induced preoplastic lesions and tumors[27-30].

In various studies, the milk products fermented by various strain of Lactobacillus, Streptococcus, and Bifidobacterium have shown potential antimutagenic activities in the

Salmonella typhimurium mutagenicity assay[31-33]. On the other hand, some strains found in buttermilk, kefir and dickmilch did not show the antimutagenic activity[34]. The protective effects were connected to the fermentation process and dependent on cell number, and the growth stage of bacteria seems to play a pivotal role in antimutagenicity. In the linear growth phase, a profound increase in antimutagenic activity occurs, reaching a maximum level of bacterial activity then decrease in the stationary growth phase[35-36].

The carcinogenic effect of endogenous toxic and genotoxic compounds is probably influenced by the activity of bacterial enzymes NAD (P)H dehydrogenase, nitroreductase, β-glucoronidase, β-glucosidase and 7-α-dehydroxylase[37]. Harmful and beneficial bacteria found in intestine differs in their enzymatic activities[38].

However, in humans, there is no evidence available on whether probiotics and prebiotics can prevent the initiation of colon cancer. Epidemiologic studies are contradictory; some studies could not find an association between the consumption of fermented-milk products and the risk of colon cancer[397-40], whereas other studies showed a lower incidence of colon cancer in persons consuming fermented- milk products or yogur[41-43].

Such anticarcinogenic properties have also been studied at a molecular level. There are fifty-seven cytochrome P450 encoded in the human genome, which is a part of large group of enzymes in human system, mainly catalyzing the metabolism of steroids, bile acids, eicosanoids, drugs and xenobiotic chemicals[44] . However, some of the P450s are also active carcinogens. Past epidemiological researches have shown increased risk of colon cancer in individuals with high P4501A2 activity. The metabolic activation of food-borne heterocyclic amines to colon carcinogens in humans is hypothesized to occur via N-oxidation followed by O-acetylation to form the N-acetoxy arylamine that binds to DNA to give carcinogen-DNA adducts. These steps are catalyzed by hepatic cytochrome P4501A2 and acetyltransferase-2 (NAT-2), respectively[45]. It has hypothesized that that probiotics such as Bifidobacterium could lower the risks of colon cancer, by producing metabolites that could affect the mixed-function of P450s and subsequently affect the conversion of azoxymethane from proximate to ultimate carcinogen[46]. These had led to the suggestion that probiotic could suppress colon cancer.

In one indigenous study, found that probiotic isolated from "idly", a famous South Indian dish could exert desmutagenicity[47] on various spice mutagens, heterocyclic amines and aflatoxins. Subsequent studies on the desmutagenicity properties of probiotic suggested that the desmutagenic substances may reside in the cellular envelope of the bacterial cell wall[48]-[50]. The response of probiotics against mutagens initiated by binding of the cell wall skeleton of probiotics on mutagens[51] and the binding of heterocyclic amines by intestinal probiotics[52]. In addition, whole cells of bifidobacteria have also been found to bind with the ultimate carcinogen methylazoxymethanol[53] and mutagen-carcinogen 3-amino-1,4-dimethyl-5H-pyrido[4,3-b] indole[54], thus physically removing it via feces and subsequently minimizing its absorption into the intestinal lumen.

Fig 3.1:- The various beneficial role of probiotics impact over many diseases.

Other preventive roles of Probiotics

Hypertension

In many clinical trials, shown that consumption of fermented milk products with various lactic acid bacteria strains may result in modest reduction in blood pressure, the link of controlling mechanism is still under investigation, however one line of research has suggested that bioactive peptides resulting from the proteolytic action of probiotic bacteria on casein (milk protein) during milk fermentation may suppress the blood pressure of hypertensive individuals[55]. Studies with hypertensive rats[56-57] and one human clinical study[58]. Two active peptides (Valine-proline-proline and isoleucine-proline-proline) isolated from fermented milk by Saccharomyces cerevisea and Lactobacillus helveticus thought to be active components. These tripeptides function as angiotensin-I- converting enzyme inhibitors and reduce bold pressure.

Urogenital infection

A healthy urinary tract of female is associated with high populations of lactobacilli (especially hydrogen peroxide- producing lactobacilli)[59-62] and a pH <5.0. this fact, has led to the theory that oral consumption or pre-existing intestinal transmitted probiotics may be useful in the treatment or prevention of urogenital infections.

Lactose intolerance

Lactose intolerance is the inability to digest lactose. Lactose is a large sugar molecule that is made up of two smaller sugars, glucose and galactose. In order for lactose to be absorbed from the intestine and into the body, it must first be split into glucose and galactose. The glucose and galactose are then absorbed by the cells lining the small intestine. The enzyme that splits lactose into glucose and galactose is called lactase, and it is located on the surface

of the cells lining the small intestine.. Lactose intolerance happens when the small intestine does not make enough of the enzyme lactase. Enzymes help the body absorb foods. Not having enough lactase is called lactase deficiency. Researchers have identified a possible genetic link to primary lactase deficiency. Some people inherit a gene from their parents that makes it likely they will develop primary lactase deficiency. This discovery may be useful in developing future genetic tests to identify people at risk for lactose intolerance.

Secondary lactase deficiency results from injury to the small intestine that occurs with severe diarrheal illness, celiac disease, Crohn's disease, or chemotherapy. This type of lactase deficiency can occur at any age but is more common in infancy. The inability of adults to digest lactose is widespread, although those deficient in lactase generally tolerate lactose better from yogurt than from milk[63-65]. Babies' bodies make this enzyme so they can digest milk, including breast milk. However, Premature babies sometimes have lactose intolerance. Children who were born at full term usually do not show signs of lactose intolerance until they are at least 3 years old. Lactose intolerance is observed mainly in people with Asian, African, Native American, or Mediterranean ancestry than it is among northern and western Europeans. In African Americans, lactose intolerance often occurs as early as age 2. Lactose intolerance is very common in adults and is not dangerous. Approximately 30 million American adults have some amount of lactose intolerance by age 20. Thus probiotics can help for intolerance of this lactose by converting it to lactic acid, thus can manage their indigestion of lactose[66].

Cholesterol

Elevated levels of certain blood lipids are a risk factor for cardiovascular disease. Animal studies have suggested the efficacy of a range of LAB are able to lower the serum cholesterol levels, presumably by breaking down bile in the gut, thus help them to excrete from feces and prevent them to reabsorb in the body.

Improving Immune function

Beside of its anti-tumerogenic effects, it may also protect host from pathogen attack by means of competitive inhibition (i.e., by competing for growth) and, plays a vital role enhancement of immune system by increasing the number of IgA-producing plasma cells , increasing or improving phagocytosis as well as increasing the proportion of T lymphocytes and Natural Killer cells[67]-[68]. Many other research studies suggested that probiotics may decrease the incidence of respiratory tract infections[69] and dental carries in children[70] as well as aid in the treatment of Helicobacter pylori infections (which cause peptic ulcers) in adults when used in combination with standard medical treatments.[71] LAB foods and supplements have been shown to be effective in the treatment and prevention of acute diarrhea; decreasing the severity and duration of rotavirus infections in children as well as antibiotic associated and travellers diarrhea in adults.[72]

References:

1. Kato I, Kobayashi S, Yokokura T & Mutai M (1981) Antitumour activity of Lactobacillus casei in mice. Gann 72, 517–523.

2. Friend BA, Farmer RE & Shahani KM (1982) Effect of feeding and intraperitoneal implantation of yogurt culture cells on Ehrlich ascites tumour. Milchwissenschaft 37, 708–710.

3. Guarner, F. & Schaafsma, G. J. (1998) Probiotics. Int. J. Food Microbiol. 39: 237–238.

4. FAO/WHO. Guidelines for the evaluation of probiotics in food. Joint FAO/WHO Working Group Report on Drafting Guidelines for the Evaluation of Probiotics in Food London, Ontario, Canada, 2002.

5. Fuller, R. Probiotics in man and animals. Journal of Applied Bacteriology, 66,265-378, 1989.

6. Liong, M.T. Probiotics: A critical review of their potential role as antihypertensives, Immune modulators, hypocholesterolemics, and perimenopausal treatments. Nutrition Reviews 2007, 65,316-328.

7. Tannock, G. W. (1998) Studies of the intestinal microflora: a prerequisite for the development of probiotics. Int. Dairy J. 8: 527–533.

8. Sanders ME. Considerations for use of probiotic bacteria to modulate human health. J Nutr. 2000;130:384S-390S.

9. Kandler O (1983): Carbohydrate metabolism in lactic acid bacteria.Antonie van Leeuwenhoek 49: 209 ± 224

10. Finegold SM, Sutter VL & Mathisen GE (1983): Normal indigenous intestinal flora. In: Hentges DJ, (ed.) Human intestinal microflora in health and disease, London, Academic Press, pp 3 ± 31

11. Kitsuoka T (1984): Taxonomy and ecology of bifidobacteria. Bifidobacteria Micro⁻ora 3: 11 ± 28.

12. Scardovi V (1986): Genus Bifidobacterium. In: Mair NS, (ed.) Bergey's manual of systematic bacteriology, New York: Williams and Wilkins,pp 1418 ± 1434.

13. Hill MJ & Drasar BS (1975): The normal colonic bacterial flora. Gut 16:318 ± 323.

14. Gorbach SL (1971): Progress in gastroenterology. Intestinal microflora.Gastroenterol 60: 1110 ± 1129.

15. Gorbach SL (1986): Function of the normal human microflora. Scand. J.Infect. Dis. Suppl. 49: 17 ± 30.

16. Brown JP (1977): Role of gut bacterial flora in nutrition and health: A review of recent advances in bacteriological techniques, metabolism,and factors affecting ⁻ora composition. CRC Crit. Rev. Food Sci. Nutr.8: 229 ± 336.

17. Stark PL & Lee A (1982): The microbial ecology of the large bowel of breast-fed and formula-fed infants during the ®rst year of life. J. Med.Microbiol. 15: 189 ± 203.

18. Greenwald,P. (1992) Colon cancer overview. Cancer, **70** (suppl.), 1206–1215.

19. Drasar, B.S.; Hill, M.J. Human Intestinal Flora, Academic Press Inc.: New York, 1974.

20. Reddy, B.S.; Hamid, R.; Rao, C.V. Effect of dietary oligofructose and inulin on Colonic preneoplastic aberrant crypt foci inhibition. Carcinogenesis, **1997**, 18, 1371–1374.

21. Ballongue J, Schumann C, Quignon P. Effects of lactulose and lactitol on colonic microflora and enzymatic activity. Scand J Gastroenterol Suppl 1997;222:41–4.

22. Bouhnik Y, Flourie B, D'Agay-Abensour L, et al. Administration of transgalacto-oligosaccharides increases fecal bifidobacteria and modifies colonic fermentation metabolism in healthy humans.J Nutr 1997;127:444–8.

23. Brown I, Warhurst M, Arcot J, Playne M, Illman RJ, Topping DL.Fecal numbers of bifidobacteria are higher in pigs fed Bifidobacterium longum with a high amylose cornstarch than with a low amylase cornstarch. J Nutr 1997;127:1822–7.

24. Campbell JM, Fahey GC Jr, Wolf BW. Selected indigestible oligosaccharides affect large bowel mass, cecal and fecal short chain fatty acids, pH and microflora in rats. J Nutr 1997;127:130–6.

25. Fuller R, Gibson GR. Modification of the intestinal microflora using probiotics and prebiotics. Scand J Gastroenterol 1997;32:28–31.

26. Salminen S, Salminen E. Lactulose, lactic acid bacteria, intestinal microecology and mucosal protection. Scand J Gastroenterol Suppl 1997;222:45–8.

27. Rowland IR, Rumney CJ, Coutts JT, Lievense LC. Effect of Bifidobacterium longum and inulin on gut bacterial metabolism and carcinogen-induced aberrant crypt foci in rats. Carcinogenesis1998;19:281–5.

28. Challa A, Rao DR, Chawan CB, Shackelford L. Bifidobacterium longum and lactulose suppress azoxymethane-induced colonic aberrant crypt foci in rats. Carcinogenesis 1997;18:517–21.

29. Goldin BR, Gualtieri LJ, Moore RP. The effect of Lactobacillus GG on the initiation and promotion of DMH-induced intestinal tumors in the rat. Nutr Cancer 1996;25:197–204

30. Goldin BR, Gorbach SL. Effect of Lactobacillus acidophilus dietary supplements on 1,2-dimethylhydrazine dihydrochloride-induced intestinal cancer in rats. J Natl Cancer Inst 1980;64:263–5.

31. Renner HW, Münzner R. The possible role of probiotics as dietary antimutagens. Mutat Res 1991;262:239–45.

LAP LAMBERT ACADEMIC PUBLISHING AG & CO. KG, DUDWELLER LANDSTR, GERMANY

32. Hosoda M, Hashimoto H, Morita H, Chiba M, Hosono A. Studies on antimutagenic effect of milk cultured with lactic acid bacteria on the Trp-P2-induced mutagenicity to TA98 strain of Salmonella typhimuriumi. J Dairy Res 1992;59:543–9.

33. Abdelali H, Cassand P, Soussotte V, Koch-Bocabeille B, Narbonne JF. Antimutagenicity of components of dairy products. Mutat Res 1995;331:133–11

34. Pool-Zobel BL, Münzner R, Holzapfel WH. Antigenotoxic properties of lactic acid bacteria in the S. typhimurium mutagenicity assay. Nutr Cancer 1993;20:261–70.

35. Surono IS, Hosono A. Antimutagenicity of milk cultured with lactic acid bacteria from Dadih against mutagenic Terasi. Milchwissenschaft 1996;51:493–7.

36. Nadathur SR, Gould SJ, Bakalinsky AT. Antimutagenicity of an acetone extract of yogurt. Mutat Res 1995;334:213–24.

37. Rowland IR. Nutrition and gut microflora metabolism. In: Rowland IR, ed. Nutrition, toxicity and cancer. Boston: CRC Press, 1991: 113–36.

38. Mital BK, Garg SK. Anticarcinogenic, hypocholesterolemic, and antagonistic activities of Lactobacillus acidophilus. Crit Rev Microbiol 1995;21:175–214.

39. Kampman E, Goldbohm RA, van den Brandt PA, van't Veer P. Fermented dairy products, calcium, and colorectal cancer in the Netherlands cohort study. Cancer Res 1994;54:3186–90.

40. Kearney J, Giovannucci E, Rimm E, et al. Calcium, vitamin D, and dairy foods and the occurrence of colon cancer in men. Am J Epidemiol 1996;143:907–17.

41. Malhotra SL. Dietary factors in a study of cancer colon from cancer registry, with special reference to the role of saliva, milk and fermented milk products and vegetable fibre. Med Hypotheses 1977;3:122–34.

42. Young, TB, Wolf DA. Case-control study of proximal and distal colon cancer and diet in Wisconsin. Int J Cancer 1988;42:167–75.

43. Peters RK, Pike MC, Garabrant D, Mack TM. Diet and colon cancer in Los Angeles County, California. Cancer Causes Control 1992;3:457–73.

44. Guengerich, F.P. Cytochromes P450, drugs, and diseases. Molecular Interventions.

2003, 3, 194- 204.

45. Lang, N.P.; Butler, M.A.; Massengill, J.; Lawson, M.; Stotts, R.C.; Maurer-Jensen, M.; Kadlubar, F.F. Rapid metabolic phenotypes for acetyltransferase and cytochrome P4501A2 and putative exposure to food-borne heterocyclic amines increase the risk for colorectal cancer or polyps. Cancer Epidemiology Biomarkers and Preventions 1994, 3, 675–682.

46. Campbell, T.C.; Hayes, J.R. The effect of quantity and quality of dietary protein on drug metabolism. Federation Proceedings 1976, 35, 2470–2474.

47. Thyagaraja, N.; Hosono, A. Antimutagenicity of lactic acid bacteria from "Idly" against foodrelated mutagens. Journal of Food Protection 1993, 56, 1061–1066

48. Singh, J.; Rivenson, A; Tomita, M.; Shimamura, S.; Ishibashi, N.; Reddy, B.S. Bifidobacterium longum, a lactic acid-producing intestinal bacterium inhibits colon cancer and modulates the intermediate biomarkers of colon carcinogenesis. Carcinogenesis, 1997, 18, 833–841

49. Sekine, K.; Ohta, J.; Onishi, M.; Tatsuki, T.; Shimokawa, Y.; Toida, T.; Kawashima, T.; Hashimoto, Y. Analysis of antitumor properties of effector cells stimulated with a cell wall preparation (WPG) of Bifidobacterium infantis. Biological and Pharmaceutical Bulletin. 1995, 18, 148–153.

50. Okawa, T.; Niibe, H.; Arai, T.; Sekiba, K.; Noda, K.; Takeuchi, S.; Hashimoto, S.; Ogawa, N. Effect of LC9018 combined with radiation therapy on carcinoma of the uterine cervix. Cancer 1993, 72, 1949–1954.

51. Zhang, X.B.; Ohta, Y. Binding of mutagens by fractions of the cell wall skeleton of lactic acid bacteria on mutagens. Journal of Dairy Science 1991, 74, 1477–1481.

52. Orrhage, K.; Sillerstrom, E.; Gustaffson, J.A.; Nord, C.E.; Rafter, J. Binding of mutagenic heterocyclic amines by intestinal and lactic acid bacteria. Mutation Research 1994, 311, 239–248.

53. Kulkarni, N.; Reddy, B.S. Inhibitory effect of Bifidobacterium longum cultures on the azoxymethane-induced aberrant crypt foci formation and fecal bacterial β-

glucuronidase.Proceedings of the Society for Experimental Biology and Medicine 1994, 207, 278–283.

54. Zhang, X.B.; Ohta, Y. Microorganisms in the gastrointestinal tract of the rat prevent absorption of the mutagen-carcinogen 3-amino- 1,4-dimethyl-5H-pyrido[4,3-b]indole. Canadian Journal of Microbiology **1993**, 39, 841–845.

55. Takano, T. (1998) Milk derived peptides and hypertension reduction. Int. Dairy J. 8: 375–381

56. Nakamura, Y., Masuda, O. & Takano, T. (1996) Decrease of tissue angiotensin-I-converting enzyme activity upon feeding sour milk in spontaneously hypertensive rats. Biosci. Biotechnol. Biochem. 60: 488–489.

57. Nakamura, Y., Yamamoto, N., Sakai, K. & Takano, T. (1995) Antihypertensive effect of sour milk and peptides isolated from it that are inhibitors to angiotensin-I-converting enzyme. J. Dairy Sci. 78: 1253–1257.

58. Hata, Y., Yamamoto, M., Ohni, M., Nakajima, K., Nakamura, Y. & Takano, T. (1996) A placebo-controlled study of the effect of sour milk on blood pressure in hypertensive subjects. Am. J. Clin. Nutr. 64: 767–771.

59. Eschenbach, D. A., Davick, P. R., Williams, B. L., Klebanoff, S. J., Young-Smith, K., Critchlow, C. M. & Holmes, K. K. (1989) Prevalence of hydrogen peroxide-producing Lactobacillus species in normal women and women with bacterial vaginosis J. Clin. Microbiol. 27: 251–256.

60. Hawes, S. E., Hillier, S. L., Benedetti, J., Stevens, C. E., Koutsky, L. A., Wolner-Hanssen, P. & Holmes, K. K. (1996) Hydrogen peroxide-producing lactobacilli and acquisition of vaginal infections. J. Infect. Dis. 174: 1058–1063.

61. Hillier, S. L., Krohn, M. A., Klebanoff, S. J. & Eschenbach, D. A. (1992) The relationship of hydrogen peroxide-producing lactobacilli to bacterial vaginosis and genital microflora in pregnant women. Obstet. Gynecol. 79: 369–373.

62. Klebanoff, S. J., Hillier, S. L., Eschenbach, D. A. & Waltersdorph, A. M. (1991) Control of the microbial flora at the vagina by H2O2 generating lactobacilli.J. Infect. Dis. 164: 94–100

63. Savaiano, D. A. & Kotz, C. (1988) Recent advances in the management of lactose intolerance. Cont. Nutr. 13: 1–4.

64. Shah, N. (1993) Effectiveness of dairy products in alleviation of lactose intolerance. Food Aust. 45: 268–271.

65. S uarez, F. L., Savaiano, D. A. & Levitt, M. D. (1995) Review article: the treatment of lactose intolerance. Aliment. Pharmacol. Ther. 9: 589–597.

66. Sanders ME. Consideration for use of probiotic bacteria to modulate human health, J. Nutr. 2000;130:384S-390S.

67. Reid G, Jass J, Sebulsky MT, McCormick JK. Potential uses of probiotics in clinical practice. Clin Microbiol Rev. 2003;16:658-72.

68. Ouwehand AC, Salminen S, Isolauri E. Probiotics: an overview of beneficial effects. Antonie Van Leeuwenhoek. 2002;82:279-89.

69. Hatakka K, Savilahti E, Ponka A, Meurman JH, Poussa T, Nase L, Saxelin M, Korpela R. Effect of long term consumption of probiotic milk on infections in children attending day care centres: double blind, randomised trial. BMJ. 2001;322:1327

70. Nase L, Hatakka K, Savilahti E, Saxelin M, Ponka A, Poussa T, Korpela R, Meurman JH. Effect of long-term consumption of a probiotic bacterium, Lactobacillus rhamnosus GG, in milk on dental caries and caries risk in children . Caries Res. 2001;35:412-20.

71. Hamilton-Miller JM. The role of probiotics in the treatment and prevention of Helicobacter pylori infection. Int J Antimicrob Agents. 2003;22:360-366

72. Cremonini F, Di Caro S, Nista EC, Bartolozzi F, Capelli G, Gasbarrini G, Gasbarrini A. Meta-analysis: the effect of probiotic administration on antibiotic-associated diarrhoea. Aliment Pharmacol Ther. 2002; 16:1461-1467.

CHAPTER-4

ENVIRONMENTAL PROTECTION USING DIRECT-FED MICROBIALS IN LIVESTOCK FEEDING

S.S. Dagar[1], S.K. Shelke[2], Monica Puniya[2], KPS Sangu[3], Devesh Gupta[3] and A.K. Puniya[1]
[1]Dairy Microbiology Division, [2]Dairy Cattle Nutrition Division, National Dairy Research Institute, Karnal, Haryana - 132 001, India, [3]Depatment of Dairy Science and Technology, Janta Vedic College, Baraut (Baghpat), UP – 250611, India

ABSTRACT

Microorganisms have been inseparably associated with humans and animals in countless beneficial ways ranging from fermentation to bioremediation. A number of genera of these live microbial groups have been used as direct-fed microbials (DFM) for human and animals. DFM products are feed supplements that contain viable naturally occurring microorganisms. Of the various fungal products that have been tested, additives containing ruminal fungi have consistently improved animal productivity. The escalating use of antibiotics and other growth stimulants in animal feeds has been of growing concern due to the potential risk of residues appearing in meat and milk. Hence, the need for a safe food supply to the consumers has instigated livestock producers to explore the safer alternatives to enhance the livestock performance. DFM is one such natural preparation that has been incorporated into livestock diets in order to accomplish the objective of improved rumen efficiency. Feeding of DFM could be the safest method of improving the rumen ecological conditions, where these efficient organisms bring favorable changes. Generally, feed additive makes it easier for the animals to digest and use their feed, thus improving the feed conversion efficiency, which is a critical factor for good weight gain in animals raised for dairy and meat production. Moreover, the DFM administration in animals has also shown to reduce methane emissions. DFM are gaining popularity day-by-day amongmicrobiologists, health professionals, biotechnologists, pharmaceutical companies, food scientists and consumers; but it is necessary that all the strains must be thoroughly studied for its beneficial properties in human or animals and have confirmed 'generally regarded as safe' status.

INDEX

INTRODUCTION

Improved animal health and performance has always remained a primary objective of people associated with livestock production. Several compounds have been used to improve animal performance either by manipulation of the rumen environment (e.g., sodium bicarbonate) or by directly altering the composition and metabolic activities of rumen microorganisms (e.g., ionophores). But, with the growing concerns towards the use of antibiotics and other growth stimulants in the animal feed industry, more emphasis has been given to increasing public awareness, disease prevention and use of other natural growth promoters like direct-fed microbials (DFM). DFM are the mono or mixed cultures of live microorganisms which when fed to the host, exert beneficial health effects by improving its gastrointestinal tract microbial balance. Aside from improving the digestibility and performance of the animal, DFM detoxify toxic compounds to modulate immune system and maintain gut peristalsis and intestinal mucosal integrity (Chaucheyras and Durand, 2010). The term DFM is different from "Probiotic" in a sense that it is only restricted to the use of "live, naturally occurring microorganisms" (Yoon and Stern, 1995; Krehbiel et al., 2003). For domestic animals like cattle and buffaloes, yeasts and aerobic fungi have been successfully used to increase growth rate and production efficiency of animals. But, now a day's use of anaerobic fungi is emphasized because of its ability to produce wide array of enzymes that can even degrade the lignified walls of plant-cells. Many factors like infections, improper food, environmental conditions and ingestion of antibiotics have been described that result in imbalance of intestinal microflora of humans or animals. For many years, studies related to supplementation of microbial feed additive in the diet for the improvement of health are under progress. Now days, there are growing evidences that DFM may be useful in managing conditions like irritable bowel syndrome, lactose intolerance, chronic liver disease, pancreatitis and even certain forms of cancers. The mechanisms suggested for the action for DFM include colonization of the lower intestine, thereby limiting the growth of any potential pathogens through 'competitive exclusion' or inhibit pathogens by lowering the pH of the intestinal lumen and by producing anti-microbial proteins (bacteriocins). This chapter will cover a number of aspects related to the type of DFM, their mode of action, environmental protection using DFM, their benefits when fed to the host etc.

Bacteria as DFM

There are many bacterial DFM that are sold for use in ruminant diets with more specific applications. Most of the DFM bacteria are lactic acid producing bacteria (LAB) with lactobacilli being the most dominant micrflora, followed by the bifidobacteria, enterococci and bacilli. Among lactobacilli, Lactobacillus acidophilus is the most common microorganisms used. Most bacterial-based DFM are probably beneficial because they have effects in the lower gut and not in the rumen. For example, Lactobacillus acidophilus produces lactic acid, which may lower the pH in small intestines. Since, acidic environments are detrimental to many pathogens (Fuller, 1977); they inhibit the growth of pathogenic microbes. Early research with DFM in ruminants focused on young milk fed calves, calves being weaned or cattle being shipped (Jenny et al., 1991). These animals were thought to be stressed and have immature microbial ecosystems in their guts (Vandevoorde et al., 1991). The diversity of bacterial species used as DFM for animals and humans is presented in Table 4.1.

LAP LAMBERT ACADEMIC PUBLISHING AG & CO. KG, DUDWELLER LANDSTR, GERMANY

Table 4.1- Microorganisms applied in DFM products

Lactobacillus species	Bifidobacterium species	Other LAB	Non-lactics
Lactobacillus acidophilus	Bifidobacterium adolescentis	Enterococcus faecalis	Bacillus cereus
Lactobacillus casei	Bifidobacterium animalis	Enterococcus. faecium	Escherichia coli
Lactobacillus crispatus	Bifidobacterium bifidum	Lactococcus lactis	Propionibacterium freudenreichii
Lactobacillus gallinarum	Bifidobacterium breve	Leuconostoc mesenteroides	
Lactobacillus gasseri	Bifidobacterium infantis	Pediococcus acidilactici	
Lactobacillus johnsonii	Bifidobacterium lactis	Sporolactobacillus inulinus	
Lactobacillus paracasei	Bifidobacterium longum	Streptococcus thermophilus	
Lactobacillus plantarum			
Lactobacillus reuteri			
Lactobacillus rhamnosus			

From Holzapfel et al., (1998)

Modes of action of bacterial DFM

In ruminants, mode of action of feeding bacterial DFM is variable, which emphasizes the need for greater understanding of underlying mechanisms. Research conducted to determine the potential mode of action of bacterial DFM has most often used the human or rodent model. Bacterial DFM have been reported to modify the balance of intestinal

LAP LAMBERT ACADEMIC PUBLISHING AG & CO. KG, DUDWELLER LANDSTR, GERMANY

microorganisms, adhere to intestinal mucosa and prevent pathogen adherence or activation, influence gut permeability, and modulate immune function are discussed below.

a) Competitive Attachment

Early research (Jones and Rutter, 1972) suggested that attachment to the intestinal wall was important for enterotoxin-producing strains of E. coli to induce diarrhea. Bacterial DFM could compete with pathogens for sites of adherence on the intestinal surface. Attachment is believed to support proliferation and reduce peristaltic removal of organisms. Abu-Tarboush et al., (1996) reported that the adherence of L. acidophilus 27SC to the GIT was confirmed in young calves; the organisms used were apparently compatible with the GIT. Adhesion is thought to be mediated either nonspecifically by physicochemical factors, or specifically by adhesive bacterial surface molecules and epithelial receptor molecules (Holzapfel et al., 1998).

b) Antibacterial Effect

Many species of lactobacilli have demonstrated inhibitory activity against pathogens. Lactobacillus acidophilus has been shown to be antagonistic toward enteropathogenic E. coli, Salmonella typhimurium, Staphylococcus aureus and Clostridium perfringens (Gilliland and Speck, 1977). Mann et al., (1980) showed that a strain of E. coli, which causes illness and death when it is the sole microbial species in young lambs, could be tolerated in the presence of lactobacilli. Hydrogen peroxide produced by lactobacilli appears to be partially responsible for the antagonistic interaction (Gilliland and Speck, 1977). Hydrogen peroxide has been demonstrated to have bactericidal activity in vitro (Reiter et al., 1980); however, it might not have much involvement in the gut since oxygen is necessary for its formation by lactobacilli. A number of reports suggest that antimicrobial proteins and/or bacteriocins either mediate or facilitate antagonism by L. acidophilus (Gilliland and Speck, 1977; Barefoot and Klaenhammer, 1983). However, because of the presence of proteolytic enzymes, their importance might be limited.

c) Immune Response

Modulation of host immunity may represent another mechanism by which DFM promote intestinal health and overall well-being of the host (Erickson and Hubbard, 2000; Isolauri et al., 2001). Bacterial DFM have been shown to affect the innate, humoral and cellular arms of the immune system. Oral administration of lactobacilli generally result in an augmentation of innate immune responses (i.e., enhanced phagocytosis and natural killer cell activity), as well as an elevate production of immunoglobulin (IgA) and a decrease IgE production in both humans and animals (Erickson and Hubbard, 2000; Isolauri et al., 2001). However, influence of DFM on cytokine production and T and B cell responses show mixed results depending on the strain, dose and duration of feeding DFM, as well as the type of tissues and cells analyzed. Furthermore, some species of probiotics appear to be capable of altering the immunomodulatory effects exerted by other species. For example, L. reuteri DSM12246 was shown to potentially suppress L. casei induced production of IL-6, IL- 12, and TNF-α in dendritic cells (Christensen et al., 2002), suggesting that the composition of bacterial DFM administered should be considered. These data provide evidence that bacterial DFM have the potential to protect animals and humans against pathogenic organisms. Several mechanisms are likely involved, but an ability to adhere to and colonize the GIT is most likely important. Bacterial DFM also show promise as immune modulators, although, more research is needed to determine the underlying mechanisms.

In short, followings are the general modes of action for DFM in ruminants for improvement in their productivity.

- Production of antibacterial compounds (acids, bacteriocins, antibiotics)
- Competition with undesirable organisms for colonization space and/or nutrients (competitive exclusion)
- Production of nutrients (e.g. amino acids, vitamins) or other growth factors stimulatory to other microorganisms in the digestive tract
- Production and/or stimulation of enzymes

- Metabolism and/or detoxification of undesirable compounds
- Stimulation of immune response in host animal
- Production of nutrients (e.g. amino acids, vitamins) or other growth factors stimulatory to the host animal

Effect of bacterial DFM on animal performance

Preruminant calves

Generally, the importance of feeding DFM to neonatal livestock has been to establish and maintain normal intestinal microorganisms rather than as a production stimulant. In the neonate, the microbial population of the gastrointestinal tract (GIT) is in transition and extremely sensitive. Abrupt environmental or dietary changes may cause shifts in the microbial population of the GIT which often leads to an increased incidence of diarrhea in calves (Sadine, 1979). Gastrointestinal disorders, including diarrhea, are one of the leading causes of mortality and morbidity in neonatal calves. In terms of ruminant production systems, the efficacy of bacterial DFM has been studied most extensively in the neonatal dairy calf. Bacterial DFM, such as species of Lactobacillus, Enterococcus, Streptococcus, and Bifidobacterium have been studied in young calves and the data have been reviewed. In general, the importance of bacterial DFM (primarily Lactobacillus species) fed to young and/or stressed calves has been to establish and maintain "normal" intestinal microorganisms rather than as a production (i.e., gain and efficiency) stimulant. For dairy calves, rapid adaptation to solid feed by accelerating the establishment of ruminal and intestinal microorganisms and avoiding the establishment of enteropathogens, which often results in diarrhea, is the primary goal. In the neonate and in stressed calves, the microbial population is in transition and extremely sensitive; abrupt changes in diet or the environment can cause alterations in microbial populations in the gastrointestinal tract (Savage, 1977). Moreover, Sandine (1979) reported that fecal counts of lactobacilli are higher than coliforms in healthy animals and reverse in those suffering from diarrhea.

Feeding calves viable cultures of species of Lactobacillus and Streptococcus has been reported to decrease the incidence of diarrhea (Ewaschuk et al., 2004; Hossaini et al., 2010;

Riddell et al., 2010). In addition to decreasing the occurrence of diarrhea, some studies have indicated that inclusion of probiotics in the diet improves weight gain, feed efficiency and feed intake (Timmerman et al., 2005; Adams et al., 2008). In a more recent experiment by Hossaini et al., (2010), calves fed DFM containing L. acidophilus, L. casei, B. thennophilus, S. faecium had a significantly lower scour compared with calves fed the control diet, which confirmed the beneficial effect of DFM in reducing the incidence of diarrhea in dairy calves suggested by earlier research. The decreased incidence of diarrhea might be associated with a consistently increased shedding of Lactobacillus (Gilliland et al., 1980; Jenny et al., 1991; Abu-Tarboush et al., 1996) and an inconsistent decreased shedding of coliforms (Bruce et al., 1979) in feces in response to supplements of Lactobacillus.

Performance response is likely not important early in the preruminant's life when enteric disease is most prevalent. Improved health and reduction in the incidence or severity of diarrhea, though difficult to measure for statistical analysis, is most likely a more important response. As suggested by Newman and Jacques (1995), more experiments that include detailed information about the microbial supplement, and fecal culture data from scouring experimental animals are needed to determine the usefulness of microbial supplements in neonatal claves.

Lactating Animals

Modern day intensive production systems, especially with high producing dairy cows and buffaloes involve the feeding of high levels of concentrates in order to supply sufficient nutrients to support a high level of milk production. Feeding these high levels of concentrate often lead to metabolic dysfunction and eventually rumen acidosis; especially under conditions of poor methods of feeding and/or composition of diets. The goal of the nutritionist, when implementing high concentrate feeding is to maximize performance and efficiency, while keeping digestive disturbances such as rumen acidosis within acceptable limits through good nutritional management. Theoretically, a number of approaches can be followed to control the incidences of rumen acidosis. One approach is to inhibit the growth of lactic acid producing bacteria such as S. bovis and Lactobacillus species through the use of feed supplements such as ionophores (Callaway and Martin, 1997). Another approach is to use DFM such as Megasphaera elsdenii, a lactic acid utilizer, to regulate lactic acid levels in the rumen through increasing the population of lactic acid utilizing bacteria. Experimentally, there have been several bacteria that have potential as DFM for ruminants but have not been commercialized for a number of different reasons. For example, M. elsdenii is the major lactate-utilizing organism in the rumen of adapted cattle fed high grain diets. However, when cattle are abruptly shifted from a high-forage to high concentrate diet, the numbers of M. elsdenii are often insufficient to prevent lactic acidosis. Erasmus (1992) observed an increase in milk production for a high producing group of cows when M. elsdenii NCIMB 41125 was dosed compared to the control animals. An increase in milk production was observed for high producing second lactation cows in a field trial conducted by KK Animal Nutrition (Hagg and Henning, 2007), where M. elsdenii NCIMB 41125 were dosed after calving. Aikman et al., (2008) observed an increase in milk production and feed intake for the first 21 days of lactation when M. elsdenii NCIMB 41125 were dosed compared to the control animals.

Gomez-Basauri et al., (2001) evaluated the effect of a supplement containing L. acidophilus, L. casei, E. faecium (total lactic bacteria = 109 cfu/g) and mannanoligosaccharide on DMI, milk yield, and milk component concentration. Cows fed lactic acid bacteria and mannanoligosaccharide consumed 0.42 kg less DM and produced 0.73 kg/d more milk. The authors reported that milk yields increased over time for DFM- and mannanoligosaccharide-fed cows, whereas control cows maintained constant milk yields. Other experiments have been conducted with combinations of fungal cultures and lactic acid bacteria (Komari et al., 1999; Block et al., 2000). Milk yields were increased by 1.08 and 0.90 kg/d, respectively, when lactating cows were fed S. cerevisae in combination with Lactobacillus acidophilus or 5×10^9 cfu of yeast in combination with 5 × 109 cfu of L. plantarum/ E. faecium. Propionibacteria, which convert lactic acid and glucose to acetic and propionic acid, may also be beneficial if inoculated into the rumen, because higher concentrations of ruminal propionate the energy status of the animal. These bacteria are naturally found in high numbers in the rumen of animals fed forage and medium concentrate diets. Their supplementation as DFM product increased milk fat percentage and milk yield as well as improved health of prepartum and postpartum cows (Noeck et al., 2006; Oetzel et al., 2007).

Fungi as DFM

In adult ruminants, fungal DFM have mostly been selected to target the rumen compartment, which is the main site of feed digestion. The fungal feed additives and supplements have been shown to affect the rumen fermentation patterns.

Mode of action of Fungal DFM

Several reasons for improvements in ruminal fermentation from feeding fungal DFM have been suggested. First, DFM exerts beneficial changes in activity and numbers of rumen microbes. For example, the numbers of total ruminal anaerobes and cellulolytic bacteria increases with fungal extracts. Beharka et al., (1991) reported that young calves fed an Aspergillus oryzae fermentation extract were weaned one week earlier than untreated calves and that supplementation increased the numbers of rumen bacteria and VFA concentrations. Aspergillus fermentation extracts (Chang et al., 1999) and yeast cultures (Chaucheryas et al., 1995) have also been shown to stimulate rumen fungi directly, which improved fiber digestion. Feeding Saccharomyces cerevisiae increased the number of rumen protozoa and increased NDF digestion in steers fed straw-based diets (Plata et al., 1994). Yeasts have also been shown to stimulate acetogenic bacteria in the presence of methanogens (Chaucheryas et al., 1995), which might result in more efficient ruminal fermentation.

Second, fungal DFM may also prevent the accumulation of excess lactic acid in the rumen when cattle are fed diets containing highly fermentable carbohydrates. Specifically, extracts of Aspergillus oryzae stimulated the uptake of lactic acid by the rumen lactate-utilizers Selenomonas ruminantium (Nisbet and Martin, 1991) and M. elsdenii (Waldrip and Martin, 1993) possibly by providing a source of malic acid. Increased metabolism of

LAP LAMBERT ACADEMIC PUBLISHING AG & CO. KG, DUDWELLER LANDSTR, GERMANY

lactic acid should theoretically raise ruminal pH and this may be one reason why DFM increased numbers of rumen cellulolytic bacteria and improved fiber digestion (Arambel et al., 1987). Chaucheyras et al., (1995) reported that S. cerevisiae was able to prevent the accumulation of lactic acid production by competing with S. bovis for glucose and by stimulating the uptake of lactic acid by M. elsdenii, perhaps by supplying amino acids and vitamins. In contrast, added yeasts were unable to prevent acute episodes of lactic acidosis when fermentations were challenged with a diet rich in fermentable carbohydrates (Aslan et al., 1995). Yeast may improve ruminal fermentation because they are able to scavenge excess oxygen (Newbold et al., 1996), creating a more optimal environment for rumen anaerobic bacteria. Aspergillus extracts may improve fiber digestion because they contain esterase enzymes (Varel et al., 1993).

Recently, anaerobic fungi have also been supplemented as fungal DFM to ruminant for better utilization of fibrous feeds in terms of increased feed intake, body weight gain, enhanced milk production, and thus improved animal productivity (Dey et al., 2004; Thareja et al., 2006). Anaerobic fungi are the normal inhabitants of the rumen ecosystem. The fungi colonize the fibrous plant fragments in rumen and penetrate plant tissues making more room for bacterial attack and thus increase the area susceptible to enzymatic attack. These properties of anaerobic fungi are suggestive of manipulation of fungal numbers for better utilization of fibrous feeds.

Effect of fungal DFM on animal performance

There have been numerous studies reporting positive effects of S. cerevisiae and A. oryzae on intake and milk production of lactating cows. Supplementing diets with S. cerevisiae was shown to increase total dry matter intake, total volatile fatty acids (VFA) and propionic acid production, besides higher propionate concentration and decreased acetate to propionate ratio were determined in some experiments (Schingoethe et al., 2004; Ondarza et al., 2010; Cakiroglu et al., 2010). Higher VFA, especially propionic acid are important in terms of enhanced lactose production, milk volume and overall energy balance (Miller-Webster et al., 2009). Erasmus et al., (1992) suggested that supplementation of S.

cerevisiae tended to increase microbial protein synthesis in dairy cows and significantly altered the amino acid profile of the duodenal digesta. Wohlt et al., (1991) suggested that supplementing yeast culture before parturition and extending through peak lactation was necessary to evaluate the effect on lactating cows. Some field reports indicate increased DMI and milk production when yeast was fed during periods of heat stress, possibly reflecting the role in aiding appetite during time of stress (Huber, 1998). In beef cattle the addition of *S. cerevisiae* lead to an increase of live weight by 7.5% depending on the type of diet tested. Improvement can reach 13% in feedlot conditions, with diets rich in starch and sugars. Wallace and Newbold (1993) reported that responses recorded in trials in beef cattle tended to be higher with corn silage rather than with grass silage. In dairy cows, an improvement by around 4% of the milk yield, often associated with increased feed intake was generally reported and response was greater in early as opposed to mid or late lactation (Ali-Haimoud-Lekhal et al., 1999). *A. oryzae* in diets of lactating cows increased milk production, feed efficiency and tolerance to heat stress in some (Gomez-Alarcon et al., 1990) but not all (Higginbotham et al., 1993; Yu et al., 1997) studies.

Among microbial additives, there are evidences of definite positive relationships between anaerobic fungi in the rumen and the increased voluntary intake of low digestible fibrous feeds (McAllister et al., 1994; Ha et al., 1994; Dey et al., 2004; Thareja et al., 2006). The anaerobic fungi have been isolated from animals of different parts of the world providing evidence to suggest that they may have an important role in the digestion of fibrous materials in the rumen (Trinci et al., 1994; Tripathi et al., 2007b) through substantial colonization of plant material (Edwards et al., 2008). Different fungal species improved digestibility of dry matter and cell wall constituents of cereal straws (Manikumar et al., 2004) as well as sugarcane bagasse (Shelke et al., 2009) in the in vitro system. Incorporation of fungus increased growth rate, rumen fermentation, nutrient digestibility and nitrogen retention in sheep (Ha et al., 1994), crossbred calves (Dey et al., 2004), and buffalo calves (Sehgal et al., 2008). Tripathi et al., (2007a; 2007b) found that

LAP LAMBERT ACADEMIC PUBLISHING AG & CO. KG, DUDWELLER LANDSTR, GERMANY

administration of Piromyces sp. increased the growth rate, feed efficiency and nutritive value of wheat straw based ration in buffalo calves.

Environmental protection using DFM

Rumen is a highly diverse and complex ecosystem comprising of different microbial groups (i.e. anaerobic bacteria, fungi and protozoa) including archaebacteria (i.e. methanogens) that utilize a major part (6 to 15%) of ruminant's energy in methane production. Methane being the second most potent green house gas, lead to the global warming and poses a number of threats to the environment. Thus, the consequences of methanogenesis in the rumen is not only associated with low animal efficiency but also have a negative impact on the sustainability of their production. Since, the enteric fermentation emission is one of the major sources of methane, therefore, a number of experiments have been conducted using various compounds like halogenated methane analogues, antibiotics, bacteriocins, propionate enhancers, acetogens, methane oxidizers etc. for mitigating methane emissions. Antibiotics such as monensin have been used widely to increase animal performance and decrease enteric methane emission. However, appearance of antibiotic-resistant bacteria restricts its convenient use. Moreover, the antibiotics excreted to manures without being absorbed have been scattered on the environment (Mwenya, 2006). The alternative to antibiotics is the use of DFM that include lactic acid bacteria and yeasts as they are also found to reduce methane emission (Kalmakoff et al., 1996; Teather and Forster, 1998; Klieve and Hegarty, 1999) and acetate: propionate ratio (Martin and Nisbet, 1992; Gamo et al., 2002; Lila et al., 2004). Hydrogen, which is released in rumen during fibre degradation by cellulolytic microbes like bacteria and anaerobic fungi, is rapidly utilized by methanogens for its conversion to methane. On the other hand, acetogenic bacteria are also able to utilize hydrogen for acetate production; but their number is less in the rumen of adults. Therefore, the acetogenic bacteria could be potentially used to compete with methanogens for hydrogen utilization; thereby also preventing the energy loss occurring as a result of methane production. Chaucheyras et al., (1995) studied the effect of a live strain of S. cerevisiae on hydrogen utilization and acetate

LAP LAMBERT ACADEMIC PUBLISHING AG & CO. KG, DUDWELLER LANDSTR, GERMANY

and methane production by an acetogen and a methanogen. They concluded that the addition of yeast cells enhanced the acetogenesis of the acetogenic strain by more than fivefold, while in absence of yeasts, hydrogen was principally used for methane synthesis. Therefore, the use of yeasts as ruminant feed additives could help reducing methane, increasing rumen metabolism and hence, promoting ruminant performance and animal health. Lopez et al., (1999) also found that acetogens depress methane production when added to rumen fluid in vitro and suggested that even if a stable population of acetogens could not be established in the rumen, it might be possible to achieve the same metabolic activity using the acetogens as a daily fed feed additive. In addition, methane oxidisers can also be used as DFM. The oxidation reaction competes with the production of methane, which is a strictly anaerobic process. Methane oxidisers from gut and non-gut sources could be screened for their activity in rumen fluid in vitro and then selected methane oxidisers could be introduced into the rumen on a daily basis.

Practical applications of DFM

There are varieties of DFMs such as powder, paste, gel, and capsules available commercially. These different forms may be mixed in feed, top-dressed, given as a paste, or mixed into the drinking water or milk replacer. However, their use must be managed effectively as viability of organism can be largely affected on interactions with chlorine, water, temperature, minerals, flow rate, and antibiotics. Bacterial DFM pastes are formulated with vegetable oil and inert gelling ingredients. Non-hydroscopic whey is generally used as a carrier for bacteria based DFM. Fungal DFM products are formulated with grain by-products as carriers. Some DFM are developed for one-time dosing while others are developed for feeding on a daily basis. Most DFMs contain live bacteria; however, some contain only bacterial or fungal extracts or fermentation by-products. The best response can be observed during the following situations: (a) when a newborn animal acquire beneficial bacteria from environment, (b) during weaning or dietary changes, (c) periods of stress i.e. shipping, vaccination, and other situations , and (d) antibiotic therapy.

The stability of DFMs is crucial because the microbes must be delivered live to the animal to be effective. For this, most DFMs require storage in a cool and dry area, away from heat, direct sunlight, and high levels of humidity. They must not only survive during processing and storage but also in the gut environment. The metabolites present in culture extracts have been suggested to be the "active" ingredients.

CONCLUSION

For ruminants, DFM have been used successfully for improving rumen and gastro-intestinal health, enhancing milk production, feed efficiency and daily gain in animals.

The increased awareness and concerns among consumers over antibiotic residues in animal products and the threat of bacterial antibiotic resistance in environment has lead to an increased interest in antibiotic alternatives like DFM. On the other hand, methanogenesis, which accounts for significant loss of ruminant's energy and increased green house gases in environment, is also a major concern in present scenario. Therefore, the use of DFM for improving production efficiency without compromising animal health and environmental sustainability is most advocated.

REFERENCES

1 Abu-Tarboush,H. M., Al-Saiady,M. Y. & Keir El-Din,A. H. (1996). Evaluation of diet containing lactobacilli on performance, fecal coliform, and lactobacilli of young dairy calves. Anim. Feed Sci. Technol.,57, 39–49.

2. Adams,M. C., Luo,J., Rayward,D., King,S., Gibson,R., Moghaddam,G. H. (2008). Selection of a novel direct-fed microbial to enhance weight gain in intensively reared calves. Animal Feed Sci and Tech.,145, 41-52.

3. Aikman,P. C., Henning,P. H., Jones,A. K., Potteron,S., Siviter,J., Carter,S., Hill,S., Kirton,P. & Szoka,R. (2008). Effect of administration of Megasphaera elsdenii NCIMB 41125 lactate utilising bacteria in early lactation on the production, health

and rumen environment of highly productive dairy cows fed a high concentrate diet. KK Animal Nutrition Internal Report.

4. Ali-Haimoud-Lekhal,D., Lescoat,P., Bayourthe,C. & Moncoulon,R. (1999). Effect of Saccharomyces cerevisiae and Aspergillus orizae on milk yield and composition in dairy cows: A review. Renc. Rech. Ruminants,6, 157.

5. Arambel,M. J., Weidmeier,R. D. & Walters,J. L. (1987). Influence of donor animal adaptation to added yeast culture and/or Aspergillus oryzae fermentation extract on in vitro rumen fermentation. Nutr. Repts. Intl.,35, 433- 437.

6. Aslan,V. S., Thamsborg,M., Jorgensen,R. J. & Basse,A. (1995). Induced acute ruminal acidosis in goats treated with yeast (Saccharomyces cerevisiae) and bicarbonate. Acta. Vet. Scand.,36, 65-68.

7. Barefoot,S. F., & Klaenhammer,T. R. (1983). Detection and activity of lactacin B, a bacterioncin produced by Lactobacillus acidophilus. Appl. Environ. Microbiol.,45, 1808–1815.

8. Beharka,A. A., Nagaraja,T. G. & Morrill,J. L. (1991). Performance and ruminal development of young calves fed diets with Aspergillus oryzae fermentation extracts. J. Dairy Sci.,74, 4326-4336.

9. Block,E., Nocek,J. E., Kautz,W. P. & Leedle,J. A. Z. (2000). Direct fed microbial and anionic salt supplementation to dairy cows fed 21 days pre- to 70 days postpartum. J. Anim. Sci.,78, 304.

10. Bruce,B. B., Gilliland,S. E., Bush,L. J. & Staley,T. E. (1979). Influence of feeding cells of Lactobacillus acidophilus on the fecal flora of young dairy calves. Oklahoma Anim. Sci. Res. Rep, 207. Stillwater, OK.

11. Cakiroglu,D., Meral,Y., Pekmezci,D. & Akdag,F. (2010). Effects of live yeast culture (Saccharomyces cerevisiae) on milk production and blood lipid levels of cows in early lactation. J Animl Vet Adv.,9, 1370-1374.

LAP LAMBERT ACADEMIC PUBLISHING AG & CO. KG, DUDWELLER LANDSTR, GERMANY

12. Callaway,T.R. & Martin,S.A. (1997). Effects of cellobiose and monensin on in vitro fermentation of organic acids by mixed ruminal bacteria. J. Dairy Sci.,80, 1126-1135.

13. Chang,J. S., Harper,E. M. &. Calza,R. E. (1999). Fermentation extract effects on the morphology and metabolism of the rumen fungus Neocallimastix frontalis EB188. J. Appl. Microbiol.,86, 389-398.

14. Chaucheyras-Durand,F. & Durand,H. (2010). Probiotics in animal nutrition and health. Beneficial Microbes, 1, 3-9.

15. Christensen,H. R., Frokiaer,H & Pestka,J. J. (2002). Lactobacilli differentially modulate expression of cytokines and maturation surface markers in murine dendritic cells. J. Immunol.,168, 171–178.

16. Dey,A., Sehgal,J. P., Puniya,A. K. & Singh,K. (2004). Influence of an anaerobic fungal culture (Orpinomyces sp) administration on growth rate ruminal fermentation and nutrient digestion in calves. Asian-Aust. J. Anim. Sci.,17, 820-824.

17. Edwards,J. E., Kingston-Smith,A. H., Jimenez,H. R., Huws,S. A., Skot,K. P., Griffith,G. W., McEwan,N. R., & Theodorou,M. K. (2008). Dynamics of initial colonization of nonconserved perennial ryegrass by anaerobic fungi in the bovine rumen. FEMS Microbiol. Ecol.,66(3), 537-545.

18. Erasmus,L. J., Botha, P.M. & Kistner,A. (1992). Effect of yeast culture supplement on production, rumen fermentation and duodenal nitrogen flow in dairy cows. J. Dairy Sci.,75, 3056-3065

19. Erickson,K. L., & Hubbard,N. E. (2000). Probiotic immunomodulation in health and disease. Amer. Soc. Nutr. Sci., 403S–490S.

20. Ewaschuk,J. B., Naylor, J. M., Chirino-Trejo,M., Zello,G. A. (2004). Lactobacillus rhamnosus strain GG is a potential probiotic for calves. Can J Vet Res.,68(4), 249-53.

21. Fuller,R. (1977). The importance of lactobacilli in maintaining normal microbial balance in the crop. Br. Poult. Sci.,18, 85–94.

22. Fuller,R. (1989). Probiotics in man and animals. J. Appl. Bact.,66, 365-378.

23. Gamo, Y., Mii, M., Zhou, X. G., Sar, C., Santoso, B., Arai, I., Kimura, K. & Takahashi, J. (2002). Effects of lactic acid bacteria, yeasts and galactooligosaccharide supplementation on in vitro rumen methane production. In: J. Takahashi, & B. A. Young (Eds.), Greenhouse Gases and Animinal Agriculture (pp. 201–204). Amsterdam, The Netherlands:Elsevier Science BV.

24. Gilliland,S. E., & Speck,M. L. (1977). Antagonistic action of Lactobacillus acidophilus toward intestinal and food borne pathogens in associative cultures. J. Food Prot.,40, 820–823.

25. Gilliland,S. E., Bruce,B. B., Bush,L.J. & Staley,T. E. (1980). Comparison of two strains of Lactobacillus acidophilus as dietary adjuncts for young calves. J. Dairy Sci.,63, 964–972.

26. Gomez-Alarcon,R. A., Dudas,C. & Huber.J. T. (1990). Influence of Aspergillus oryzae on rumen and total tract digestion of dietary components. J. Dairy Sci.,73, 703–710.

27. Gomez-Basauri,J., de Ondarza,M. B. & Siciliano-Jones,J. (2001). Intake and milk production of dairy cows fed lactic acid bacteria and mannanoligosaccharide. J. Dairy Sci.,84(Suppl. 1), 283.

28. Ha,J. K., Lee,S. S., Kim,C. H., Choi,Y. J. & Min,H. K. (1994). Effect of fungal inoculation on ruminal fermentation characteristics enzyme activities and nutrient-digestion in sheep. Proc. Soc. Nutr. Physiol.,3, 197.

29. Hagg,F. M. & Henning,P. H. (2007). Evaluation of supplementation with Megasphaera elsdenii NCIMB 41125, a lactate utilizing rumen microorganism, on performance in Holstein dairy cows. KK Animal Nutrition Internal Report.

30. Higginbotham,G. E., Bath,D. L. & Butler,L. J. (1993). Effect of feeding Aspergillus oryzae extract on milk production and related responses in a commercial dairy herd. J. Dairy Sci.,76, 1484–1489.

31. Holzapfel,W. H., Haberer,P., Snel,J., Schillinger,U., & Huis in't Veld,J. H. J. (1998). Overview of gut flora and probiotics. Int. J. Food Microbiol.,41, 85–101.

32. Hossaini,S. M. R., Bojarpour,M., Mamouei,M., Asadian,A. & Fayazi,J. (2010). Effects of probiotics and antibiotic supplementation in daily milk intake of newborn calves on feed intake body weight gain, fecal scores and health condition. J Animal and Vet Adv.,9, 872-875.

33. Huber,J. T. (1998). Yeast products help cattle handle heat. Hoard's Dairyma, 143:367.

34. Isolauri,E., Sutas,Y., Kankaanpaa,Y. P., Arvilommi, H. & Salminen,S (2001). Probiotics: Effects on immunity. Am. J. Clin. Nutr.,73(Suppl. 2), 444S–450S.

35. Jenny,B. F., Vandijk,H. J. & Collins,J. A. (1991). Performance and fecal flora of calves fed a Bacillus subtilis concentrate. J. Dairy Sci.,74, 1968–1973.

36. Jones,G. W. & Rutter,J. M. (1972). Role of K88 antigen in the pathogenesis of neonatal diarrhoea caused by Escherichia coli in piglets. Infect. Immun.,6, 918–927.

37. Kalmakoff,M. L., Barlett,F. & Teather,R. M. (1996). Are ruminal bacteria armed with bacteriocin? J Dairy Sci, 79, 2297–2306.

38. Klieve,A. V. & Hegarty,R. S. (1999). Opportunities for biological control of methanogenesis. Aust J Agric Res, 50, 1315–1319.

39. Komari,R. K., Reddy,Y. K. L., Suresh,J. & Raj,D. N. (1999). Effect of feeding yeast culture (Saccharomyces cerevisae) and Lactobacillus acidophilus on production performance of crossbred dairy cows. J. Dairy Sci.,82(Suppl. 1), 128.

40. Krehbiel,C. R., Rust,S. R., Zhang,G. & Gilliland,S. E. (2003). Bacterial direct-fed microbials in ruminant diets: Performance response and mode of action. J. Anim Sci., 81,120-132.

41. Lila,Z. A., Mohammed,N., Yasui,T., Kurokawa,Y., Kanda,S. & Itabashi,H. (2004). Effects of twin strain of Saccharomyces cerevisiae live cells on mixed ruminal microorganism fermentation in vitro. J Anim Sci, 82, 1847–1854.

42. Lopez,S., Valdes,C., Newbold,C. J. & Wallace,R. J. (1999). Influence of sodium fumarate on rumen fermentation in vitro. British journal of Nutrition, 81, 59-64.

43. Manikumar,B., Puniya,A. K., Singh,K. & Sehgal, J. P. (2004). In vitro degradation of cell wall and digestibility of cereal straws treated with anaerobic ruminal fungi. Indian J. Exp. Biol.,42, 636-638.

44. Mann,S. O., Grant,C. & Hobson,P. N. (1980). Interactions of E. coli and lactobacilli in gnotobiotic lambs. Microbios Lett.,15, 141–144.

45. Martin,S. A. & Nisbet,D. J. (1992). Effect of Direct-Fed Microbials on Rumen Microbial Fermentation. J Dairy Sci, 75, 1736-1744.

46. McAllister,T. A., Bae,H. D., Yanke,L. J., Cheng,K. J. & Muir,A. (1994). Effect of condensed tannins from birdsfoot trefoil on endoglucanase activity and the digestion of cellulose filter paper by ruminal fungi. Can. J. Microbiol.,40, 298-305.

47. Miller-Webster,T., Hoover,W. H., Holt,M., Nocek,J. E. (2009). Influence of yeast culture on ruminal microbial metabolism in continuous culture. J. Dairy Sci.,85, 2014- 2021.

48. Mwenya,B., Sar,C., Pen,B., Morikawa,R., Takaura,K., Kogawa,S., Kimura,K., Umetsu,K. & Takahashi,J. (2006). Effect of feed additives on ruminal methanogenesis and anaerobic fermentation of manure in cows and steers. International Congress Series 1293, 209-212.

49. Newbold,C. J., Wallace,R. J. & McIntosh,F. M. (1996). Mode of action of the yeast Saccharomyces cerevisiae as a feed additive for ruminants. Brit. J. Nutr.,76, 249.

50. Nisbet,D. J., & Martin,S. A. (1991). Effect of a Saccharomyces cerevisiae culture on lactate utilization by the ruminal bacterium Selenomonas ruminantium. J. Anim. Sci.,69, 4628.

LAP LAMBERT ACADEMIC PUBLISHING AG & CO. KG, DUDWELLER LANDSTR, GERMANY

51. Nocek,J. E., & Kautz,W. P. (2006). Direct-fed microbial supplementation on ruminal digestion, health, and performance of pre- and postpartum dairy cattle. J. Dairy Sci.,89, 260–266.

52. Oetzel,G. R., Emery,K. M., Kautz,W. P. & Nocek,J. E. (2007). Direct-fed microbial supplementation and health and performance of pre- and postpartum dairy cattle: A field trial. J. Dairy Sci.,90, 2058–2068.

53. Ondarza,de M. B., Sniffen,C. J., Graham,H. & Wilcock,P. (2010). Case study: Effect of supplemental live yeast on yield of milk and milk components in high-producing multiparous Holstein cows. Professional Animl Scientist.,26, 443-449.

54. Plata,F. P., Mendoza,G. D., Barcena-Gama,J. R. & Gonzalez,S. M. (1994). Effect of a yeast culture (Saccharomyces cerevisiae) on neutral detergent fiber digestion in steers fed oat straw based diets. Anim. Feed Sci. Tech.,49, 203-210.

55. Reiter,B., Marshall,V. M. & Philips,S. M. (1980). The antibiotic activity of the lactoperocidase-thiocyanate-hydrogen peroxide system in the calf abomasum. Res. Vet. Sci.,28, 116–122.

56. Riddell,J. B., Gallego,A. J., Harmon,D. L. & McLeod,K. R. (2010). Addition of a Bacillus based probiotic to the diet of preruminant calves: Influence on growth, health, and blood parameters. Intern J Appl Res Vet Med.,8, 78-85.

57. Sadine,W.E. (1979). Roles of lactobacillus in the intestinal tract. J. Food. Prod.,42, 259-262.

58. Savage,D. C. (1977). Microbial ecology of the gastrointestinal tract. Annu. Rev. Microbiol.,31, 107–133.

59. Schingoethe,D. J., Linke,K. N., Kalscheur,K. F. & Hippen,A. R. (2004). Feed efficiency of mid-lactation dairy cows fed yeast culture during summer. J. Dairy Sci.,87, 4178–4181.

60. Sehgal,J. P., Jit,D., Puniya,A. K. & Singh,K. (2008). Influence of anaerobic fungal administration on growth, rumen fermentation and nutrient digestion in female buffalo calves. J. Anim. Feed Sci.,17, 510–518.

61. Shelke,S. K., Aruna Chhabra, Puniya,A. K. & Sehgal,J. P. (2009). In vitro degradation of sugarcane bagasse based ruminant rations using anaerobic fungi. Annals of Microbiol.,59(3), 415-418.

62. Teather,R. M. & Froster,R. J. (1998) Manipulating the rumen microflora with bacteriocin to improve ruminant production. Can J Anim Sci, 78, 57–69

63. Thareja,A., Puniya,A. K., Goel,G., Nagpal,R., Sehgal,J. P., Singh,P. & Singh,K. (2006). In vitro degradation of wheat straw by anaerobic fungi from small ruminants. Arch. Anim. Nutr.,60, 412-417.

64. Timmerman,H. M., Mulder,L., Everts,H., van Espen,D. C., van der Wal,E., Klaassen,G., Rouwers,S. M. G., Hartemink,R., Rombouts,F. M., Beynen,A. C. (2005). Health and growth of veal calves fed milk replacers with or without probiotics. J Dairy Sci.,88, 2154- 2165.

65. Trinci,A. P. J., Davies,D. R., Gull,K., Lawrence,M. I., Nielsen,B. B., Rickers,A. & Theodorou,M. K. (1994). Anaerobic fungi in herbivorous animals. Mycol. Res.,98, 129–152.

66. Tripathi,V. K., Sehgal,J. P., Puniya,A. K. & Singh,K. (2007a). Hydrolytic activities of anaerobic fungi isolated from wild blue bull (Boselaphus tragocamelus). Anaerobe., 13, 36-39.

67. Tripathi,V. K., Sehgal,J. P., Puniya,A. K. & Singh,K. (2007b). Effect of administration of anaerobic fungi isolated from cattle and wild blue bull (Boselaphus tragocamelus) on growth rate and fibre utilization in buffalo calves. Arch. Anim. Nutr.,61, 416-423.

68. Vandevoorde,L., Christianens,H. & Verstraete,W. (1991). In vitro appraisal of the probiotic value of intestinal lactobacilli. World. J. Microbiol. Biotechnol.,7, 587-592.

69. Varel,V. H., Kreikemeier,K. K., Jung, H.J.G. & Hatfield,R. D. (1993). In vitro stimulation of forage fiber degradation by ruminal microorganisms with Aspergillus oryzae fermentation extract. Appl. Environ. Microbiol.,59, 3171-3176.

LAP LAMBERT ACADEMIC PUBLISHING AG & CO. KG, DUDWELLER LANDSTR, GERMANY

70. Waldrip,H. M., & Martin,S. A. (1993). Effects of an Aspergillus oryzae fermentation extract and other factors on lactate utilization by the ruminal bacterium Megasphaera elsdenii. J. Anim. Sci.,71, 2770-2776.

71. Wallace, R. J. & Newbold, C. J. (1993). Rumen fermentation and its manipulation: The development of yeast culture as feed additives. In: T. P. Lyons (ed.), Biotechnology in the Feed Industry (pp. 173-192). Kentucky, Alltech Technical Publications.

72. Wohlt,J. E., Finkelstein,A. D. & Chung.C. H. (1991). Yeast culture to improve intake, nutrient digestibility, and performance by dairy cattle during early lactation. J. Dairy Sci.,74, 1395–1400.

73. Yoon,I. K., & Stern,M. D. (1995). Influence of direct-fed microbials on ruminal microbial fermentation and performance of ruminants: A review. Asian-Australas. J. Anim. Sci.,8, 533–555.

74. Yu,P., Huber,J. T., Theurer,C. B., Chen,K. H., Nussio,L. G. & Wu,Z. (1997). Effect of steam-flaked or steam-rolled corn with or without Aspergillus oryzae in the diet on performance of dairy cows fed during hot weather. J. Dairy Sci.,80, 3293–3297.

CHAPTER-5

Microbial Degradation of Lignin for Remediation and Decolorization of Industrial Effluent Dyes

Ira Chaudhary, Smita Rastogi and Neelam Pathak
Department of Biotechnology, Integral University
Lucknow-226026, UP-India

ABSTRACT

Lignin is the second most abundant natural aromatic polymer after cellulose in terrestrial ecosystems and represents nearly 30% of the organic carbon sequestered in the biosphere and it protects cellulose towards hydrolytic attack by saprophytic and pathogenic microbes. It is a three dimensional natural plant biopolymer formed by radical coupling of hydroxycinnamyl subunits called monolignols mainly p-coumaryl, sinapyl and coniferyl alcohols. The ligninolytic microbes exhibit a unique strategy for lignin degradation, which is based on unspecific one-electron oxidation of the benzenic rings catalyzed by synergistic action of extracellular haemperoxidases and peroxide-generating oxidases.

The lignin degrading enzymes play an important role in the treatment and decolorization of a wide spectrum aromatic dye from industrial effluent. It can catalyze degradation of aromatic dyes either by precipitation or by opening the aromatic ring structure. Lignin degrading enzymes can be easily employed for the remediation of commercial dyes. Lignin degrading enzymes either soluble or immobilized have been effectively exploited in batch as well as in continuous processes for the treatment of synthetic dyes present in industrial effluents at large scale. However, recalcitrant dyes were also decolorized by the action of lignin degraders in the presence of redox mediators.

The contamination of water by dye-containing effluents is of environmental concern. Due to the increasing awareness and concern of the global community over the discharge of synthetic dyes into the environment and their persistence, much attention has been focused on the remediation of these pollutants. Among the current pollution control technologies, biodegradation of synthetic dyes by different lignin degrading microbes is emerging as an

LAP LAMBERT ACADEMIC PUBLISHING AG & CO. KG, DUDWELLER LANDSTR, GERMANY

effective and promising approach. The bioremediation potentials of many microbes for synthetic dyes have been demonstrated and those of others to be explored in future. The biodegradation of synthetic dyes is an economic, effective, environmental as well as bio friendly process. Moreover bioremediation of xenobiotics including synthetic dyes by various lignin degrading microbes will hopefully prove a green solution to the problem of environmental soil and water pollution in future.

Keywords- Lignin, monolignols, bioremediation, xenobiotics, biodegradation, dyes

INDEX

INTRODUCTION

Lignin is the second most abundant natural aromatic polymer after cellulose in terrestrial ecosystems and represents nearly 30% of the organic carbon sequestered in the biosphere (Boudet et al. 2003). Lignins are interlinked with cellulose and hemicellulose conferring structural strength, rigidity and impermeability to the woody cell wall, while providing natural resistance against chemical or microbial attack and environmental stresses (Foster et al. 2010). Additionally, lignin waterproofs the cell wall thus enabling transport of water and solutes through the vascular system.

Lignin is an amorphous, aromatic, water insoluble, heterogeneous, three-dimensional and cross-linked polymer with low viscosity (Sjöström 1993; Brunow 2001). Lignins are complex racemic aromatic heteropolymers (Wong 2009) derived mainly from p-coumaryl, coniferyl, and sinapyl alcohols. The formation of p-hydroxyphenyl (H), guaiacyl (G), and syringyl (S) phenylpropanoid units occurs when the respective monolignols are incorporated into the lignin polymer (Karkonen and Koutaniemi 2010; Vogt 2010). The main role of lignin is the protection of cellulose and hemicellulose from the plant cell wall. Lignin also provides interesting structural properties to wood in nature like hardness, waterproofness and resistance to microbial attack, as lignin is a very recalcitrant polymer. Only selected organisms can degrade lignin, (Hammel, 1997). They used an extracellular complex of enzymes that performs different redox reactions leading lignin degradation. There are different groups of enzymes that participate in this process, although each microbial species has its own enzymatic profile (Hatakka, 1994). Basically the attack of lignin is produced by peroxidases and/or laccases (polyphenol oxidases) generating free radicals that perform the oxidation of the polymer by using hydrogen peroxide in the case of peroxidases or molecular oxygen for laccases. The microbial sources; lignin peroxidases, manganese peroxidases, vanadium haloperoxidases, versatile peroxidases, dye decolorizing peroxidases have been employed for the remediation of commercial dyes. Soluble and immobilized peroxidases have been successfully exploited in batch as well as in continuous processes for the treatment of synthetic dyes with complex aromatic molecular structures present in industrial effluents at large scale. However, recalcitrant dyes were also decolorized by the action of peroxidases in the presence of redox mediators. The present paper discusses the structure of lignin responsible for its recalcitrant nature and various biologninolytic systems (microbes and enzymes). Further the applicability of biologninolytic systems in various fields is also presented.

MICROBIAL DEGRADATION OF LIGNIN

A variety of microorganisms like bacteria, actinomycetes and fungi are known for degradation of lignin, however, the extent of degradation vary with the type of microorganisms.

Lignin degradation by fungi has been studied extensively in recent years to safeguard the commercial interests of the pulp and paper industry, as well as its biodegradation for environment restoration (Akhtar et al. 1997; Hatakka 2001; Scott and Akhtar 2001). It is reported that fungi degrades lignin more rapidly and extensively than any other microorganisms (Dix and Webster 1995). Lignin degrading fungi known so far mostly belong to class basidiomycetes and ascomycetes. However, the basidiomycetous fungi have been observed as better degraders of lignin in woods, cereals, and grasses. Among ascomycetes, Xylaris hypoxylon, X. polymorpha, Ustulina vulgaris have been reported to aid in lignin degradation.

The wood-decomposing basidiomycetes completely metabolize lignin, exhibit the highest reported rates, and are the most studied. The model fungus for lignin degradation is P. chrysosporium (Kirk 1990), but recently certain other fungi with ligninolytic activity have also been studied, some of which are Trametes versicolor, Cyathus bulleri, C. stercoreus, Phlebia radiata, Pycnoporus sanguineus, P. cinnabarinus, Pleurotus ostreatus, P. eryngii, P. sajorcaju, P. florida, Phanerochaete chrysosporium, Cereporiopsis subvermispora, Sporotrichium thermophile, Cariolus pruinosum, Dichomitus squalens, Lentinus edodes, Panus tigrinus, Rigidosporus lignosus, Oudemansiella radicata, Polyporus pletenis, P. brumalis, Phlebia tremellosus, P. ochraceofulva etc. (Falcón et al. 1995; Eggert et al. 1996; Vasdev and Kuhad 1994; Hatakka 2001). Evidence indicates that the growth substrates of these fungi are cellulose and hemicelluloses, but lignin degradation occurs at the end of primary growth by secondary metabolism in deficiency of nutrients, such as nitrogen, carbon, or sulphur. Fungal attack is an oxidative and non-specific process, which decreases methoxy, phenolic, and aliphatic content of lignin, cleaves aromatic rings, and creates new carbonyl groups. The study of degraded polymeric lignin isolated from wood partially decayed by fungi has disclosed heavy oxidation with formation of numerous carboxyl groups. These changes in the lignin molecule result in depolymerization and carbon dioxide production. The presence of oxygen is reported to stimulate lignin degradation by some fungi, and these fungi are able to mineralize up to 75% of lignin. The degradation of lignin is either selective or non-selective (i.e. simultaneous decay) (Blanchette 1995; Hatakka 2001). In selective decay (e.g. by Ceriporiopsis subvermispora, Dichomitus squalens, Phanerochaete chrysosporium, Phlebia radiata), lignin and hemicellulose are degraded significantly more than cellulose, while in non-selective decay (e.g. by Trametes versicolor and Fomes fomentarius), equal amounts of all components of lignocellulose are degraded (Blanchette 1995; Hatakka 2001). Selective lignin degradation occurs under special conditions or at the beginning of decay, and is followed by hemicellulose and cellulose degradation (Dix and Webster 1995). Some fungi, such as Ganoderma applanatum, Heterobasidion annosum, and Phellinus pini, are capable of carrying out both types of decay (Blanchette 1995). Nutritional factors may control the type of attack, and the fungus may degrade wood both selectively and simultaneously in the very same wood stem (Eaton and Hale 1993). The majority of fungi grow on hardwoods, but certain species grow on softwoods, such as Heterobasidion annosum, Phellinus pini, and Phlebia radiata (Blanchette 1995). Fungi produce various enzymes involved in lignin degradation, but also produce cellulases, xylanases, and other hemicellulases (Hatakka 2001). The ligninolytic enzymes of white-rot fungi are unspecific, and thus, these fungi are considered to be potential microorganisms for bioremediation of polluted soils (Orth et al. 1994; Paszczynski and Crawford 1995; Hatakka 2001). Almost all fungi produce MnP and Lac, but only some of them produce LiP (Hatakka 2001). Recently, it has been shown that MnP produced by the fungus Nematoloma frowardii or Phlebia radiata alone can mineralize lignin. Thus, MnP is probably the most important enzyme for lignin mineralization by fungi. Some litter-decomposing fungi produce similar enzymes (e.g. MnP) to those of wood degrading fungi, but less is known about their lignin degradation efficiency (Hofrichter 2002).

Several bacteria are also capable of degrading lignin and lignin-related aromatic compounds (Zimmermann 1990; Perestelo et al. 1996). In general, bacteria degrade wood slowly, and degradation takes place on wood surfaces with high moisture content (Blanchette et al. 1991; Blanchette 1995). Few bacterial species have been found to degrade lignified wood cells, which have also been confirmed by ultrastructural investigations. They are capable of attacking both softwood and hardwood by first colonizing the parenchyma cell walls. Wood-degrading bacteria have primarily cellulolytic

LAP LAMBERT ACADEMIC PUBLISHING AG & CO. KG, DUDWELLER LANDSTR, GERMANY

and pectinolytic activities and pure bacterial cultures are unable to perform efficient lignin degradation. However, bacteria degrade lignocellulose in mixed cultures, either in mixed bacterial cultures, or more commonly, in bacterial and fungal cultures together, and bacteria probably consume mainly fungal by-products in mixed cultures (Vicuña et al. 1993; Vicuña 2000). This is because of the lack of penetrating ability by bacteria alone. Although bacteria can directly attack fibers, vessels and tracheids, few species and strains can degrade all the cell wall components. The lignin degradation mechanism of bacteria is more specific than that of fungi, and one bacterial species is able to cleave only one type of bond in the lignin polymer, and bacteria generally cause a low percentage of mineralization of lignin from lignocellulosic materials (Vicuña et al. 1993). Wood-degrading bacteria have a wider tolerance of temperature, pH, and oxygen limitations than fungi (Daniel and Nilsson 1998). Thus, bacterial degradation can be observed when the growth of fungi is repressed in extreme environment or substrate conditions, such as wood saturated with water, oxygen limitation, high extractive content, high concentration of lignin, or wood treated with chemical preservatives (Blanchette 1995; Daniel and Nilsson 1998). This lignin degradation capability of prokaryotes is due to the presence of proteins with typical features of the multi-copper oxidase enzyme family (Claus 2003).

Another group of microbes, actinomycetes also thrive well in environments rich in lignocellulose, such as soil, compost, heaps of hay, straw, or wood chips (Lacey 1988). Actinomycetes have also been reported to mineralize lignin successfully, although not as fast or as comprehensively as fungi (Rüttimann et al. 1991). The degradation of lignin results in the release of lignin rich, water-soluble fragments called acid precipitable polymeric lignin (APPL) (Spiker et al. 1992). Polyphenolic and polymeric lignin fragments have lower molecular mass and lower methoxyl content than native lignin, and these are associated with bacterial protein or hemicellulosic carbohydrates (Zimmermann 1990). Although lignin mineralization by actinomycetes is not as efficient as by fungi, it is still more efficient than by unicellular bacteria. Some of the most active lignin degraders among actinomycetes are Streptomycetes badius, S. cyaneus, and Thermomonospora mesophila.

Actinomycetes have been claimed to produce extracellular peroxidase (Mercer et al. 1996). It is suggested that phenol oxidase produced by Streptomyces cyaneus participates in lignin degradation much more than the peroxidase produced (Berrocal et al. 2000). Laccases are also found in actinomycetes such as Streptomyces griseus and S. lavendulae (Suzuki et al. 2003). In Streptomyces cyaneus, a laccase type phenol oxidase is produced during growth under solid substrate fermentation conditions, and it is suggested that this enzyme is involved in the solubilization and mineralization of lignin from wheat straw.

In nature, lignin is probably degraded by an array of microorganisms, although abiotic degradation may also occur in special environments, such as those due to alkaline chemical spills (Blanchette et al. 1991) or UV radiation. In aqueous or other anaerobic environments, polymeric lignin is not degraded and wood may persist in undegraded form for several hundreds or thousands of years (Blanchette 1995). Slow abiotic degradation, favored by high-temperature, acidic, or alkaline environments, releases small fragments. Amongst microorganisms, a few bacteria, actinomycetes and fungi are known to degrade lignin; however, the extent of degradation varies with the type of microorganisms. The degradation of lignin has been studied extensively in recent years to safeguard the commercial interests of the pulp and paper and textile industries, increasing forage digestibility, as well as its biodegradation for environment restoration (Akhtar et al. 1997; Hatakka et al. 2001; Scott and Akhtar 2001).

LIGNIN-DEGRADATION PATHWAY

Microbial ligninolytic enzymes, viz., Lac, LiP and MnP oxidize the lignin polymer, thereby generating aromatic radicals. These evolve in different non-enzymatic reactions, including C4-ether breakdown, aromatic ring cleavage, Cα-Cβ breakdown and demethoxylation. The aromatic aldehydes released from the breakdown of Cα-Cβ in lignin, or synthesized de novo by microbes are the substrates for aryl-alcohol oxidase (AAO)-catalyzed generation of H_2O_2 in cyclic redox reactions also involving aryl-alcohol dehydrogenase (AAD). If the phenoxy radicals produced after C4-ether breakdown are not first reduced by oxidases to phenolic compounds, may repolymerize on the lignin polymer. The resulting phenolic compounds may again be reoxidized by laccases or peroxidases. Phenoxy radicals can also be subjected to Cα-Cβ breakdown, generating p-quinones. Quinones participate in oxygen activation in redox cycling reactions involving quinone reductase (QR), laccases and peroxidases. This results in reduction of the ferric iron present in wood, either by superoxide cation radical or directly by the semiquinone radicals and its reoxidation with concomitant reduction of H_2O_2 to hydroxyl free radical (OH·). Hydroxyl free radical, a very strong oxidizing agent, can initiate the attack on lignin in the initial stages of wood decay, when the small pores in the intact cell wall prevent the penetration of ligninolytic enzymes. Then, lignin degradation proceeds by oxidative attack of the enzymes described above. In the final steps, simple products resulting from lignin degradation enter the microbes and get incorporated into intracellular catabolic routes. The entire mechanism of lignin degradation is depicted in Figure 1.

LIGNINOLYTIC ENZYMES

Lignin degradation requires unspecific and extracellular enzymes because of its random structure and high molecular mass. Primarily three enzymes, i.e., laccase, manganese peroxidase, and lignin peroxidase are responsible for lignin degradation (Vasdev et al. 1995; Pointing 2001). Some wood decaying fungi produce all three enzymes while others produce either one or two of them.

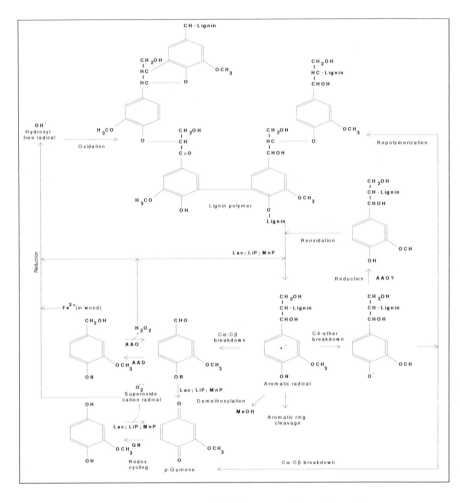

Figure 5.1- Lignin degradation pathway including enzymatic reactions and oxygen activation.

LAP LAMBERT ACADEMIC PUBLISHING AG & CO. KG, DUDWELLER LANDSTR, GERMANY

1. Laccases (Lac; EC 1.10.3.2)

Laccases are widespread in nature and are implicated in lignin degradation (Thurston 1994; Eggert et al. 1996; Alexandre and Zhulin 2000; Suzuki et al. 2003). Lac [p-benzenediol : oxygen oxidoreductase] is a blue copper-containing oxidase belonging to a family of multicopper oxidases, which also includes ascorbate oxidase, ceruloplasmin and bilirubin oxidase (Hoegger et al. 2006). Lac catalyzes the oxidation of various aromatic compounds (predominantly phenols) with the reduction of oxygen to water (Thurston 1994).

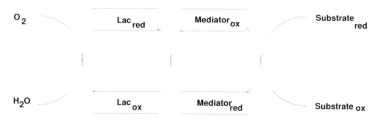

Figure 5.2. Mechanism of laccase action with reference to mediator

Copper atoms (four in number) in Lac play important role in its catalytic mechanism. Being large in size (MW 70 kDa), Lac cannot penetrate deep into wood (Bourbonnais et al. 1997). Moreover, due to its 0.5–0.8 V redox potential (low), Lac is unable to oxidize non-phenolic (C4-etherified) lignin units with >1.5 V redox potential (high) (Galli and Gentili 2004). Therefore, an oxidation mediator (i.e., a small molecule that is able to extend the effect of Lac to non-phenolic lignin units and to overcome the accessibility problem) is often used with Lac. Lac-mediated oxidation of non-phenolic lignin units can follow an electron transfer, a radical hydrogen atom transfer, or an ionic mechanism, depending on the mediator (Barreca et al. 2004). Several organic and inorganic compounds, such as thiol and phenol aromatic derivatives, N-hydroxy compounds and ferrocyanide are used as mediators (Susana and José 2006). The mechanism of Lac action with respect to mediator is shown in Figure 5.2.

2. Lignin Peroxidase (LiP; Ligninase; EC 1.11.1.14)

Lignin peroxidase (LiP; EC 1.11.1.14) is an extracellular heme containing peroxidase, which is dependent on H_2O_2, and has an unusually high redox potential and low optimum pH, typically showing little specificity towards substrates and degrades a variety of lignin related compounds (Gold and Alic 1993). Similar to MnP, LiP also oxidizes the substrate by two consecutive one-electron oxidation steps with intermediate cation radical formation (Kirk et al.,1990, Hatakka 2001). This oxidizes non-phenolic units of lignin, which is then decomposed chemically. LiP preferentially cleaves the Cβ-Cβ bond in the lignin molecule but is also capable of ring opening and other reactions. Since limited nutrient levels (mostly C or N) induce lignin oxidation, production of LiP and MnP is generally optimal at high oxygen tension but is repressed by agitation in submerged liquid culture. The mechanism of LiP is shown in Figure 3.

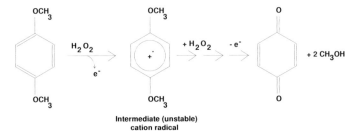

Figure 5.3. Mechanism of action of lignin peroxidase.

3. Manganese Peroxidase (MnP; EC 1.11.1.13)

The third major enzyme responsible for lignin degradation is MnP. It is an extracellular heme containing peroxidase with a high molecular weight (32 and 62.5 kDa) and secreted in multiple isoforms (Urzua et al. 1995). It was first isolated from the extracellular medium of ligninolytic cultures of white rot fungus P. chrysosporium and it is considered to be a key enzyme in ligninolysis by white-rot fungi.

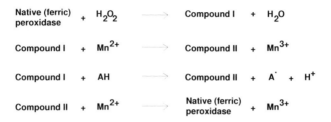

Figure 5.4. Mechanism of action of manganese peroxidase.

Similar to LiP, it oxidizes the substrate by two consecutive single electron oxidation steps with intermediate (unstable) cation radical formation. MnP requires H_2O_2 as a cosubstrate and oxidizes Mn^{2+} found in wood and soil to Mn^{3+} (Hatakka et al. 2001; Hofrichter 2002). Mn^{3+} is highly reactive, which complexed with an organic acid acts as a primary agent in ligninolysis. Mn^{3+} then oxidizes phenolic rings of lignin to unstable free radicals followed by spontaneous decomposition. MnP enzyme intermediate are analogous to other peroxidases (Wariishi et al. 1989). Thus, the native MnP is oxidized by H_2O_2 to Compound I, which can be reduced by Mn^{2+} and phenols to generate Compound II. Compound II is reduced back to a resting state by Mn^{2+}, but not by phenols. Therefore Mn^{2+} is necessary to complete the catalytic cycle and shows saturation kinetics (Pease and Tien 1992). The overall reaction catalyzed by MnP is shown in Figure 4.

The crystal structure of MnP shows similarities with LiP; the active site has a proximal His ligand hydrogen bonded to an Asp and a distal side peroxide-binding pocket consisting of a catalytic His and Arg (Sundaramoorthy et al. 1994). However, there is also a proposed manganese binding site involving Asp-179, Glu-35, Glu-39 and a heme propionate. In MnP there are five disulfide bonds.

LAP LAMBERT ACADEMIC PUBLISHING AG & CO. KG, DUDWELLER LANDSTR, GERMANY

POTENTIAL APPLICATIONS OF LIGNIN-DEGRADING ENZYMES

A number of industrial applications for lignin degrading enzymes have been proposed that include delignification and pulp bleaching, bioremediation of contaminating environmental pollutants (Schlosser et al, 1997), prevention of wine decolouration, medical applications, oxidation of dye and their precursors, enzymatic conversion of chemical intermediates, production of chemicals from lignin etc. Oxidoreductive enzymes play an important role in degradation and transformation of polymeric substances. The products which are partially degraded or oxidized by the microbial cells are completely mineralized. Lignin-degrading enzymes are one such group of oxidoreductive enzymes, which have practical application in dye decolorization of effluent water and bioremediation of polluted environment.

1. Dye decolorization

Enzyme activity can inhibit by the decolorization, textile effluent not only dyes but also salts, sometimes at very high ionic strength and extreme pH values, chelating agents, precursors, by-products and surfactants (Abadulla et al. 2000). In industrial decolorization processes chromophore compounds such as azo, triarylmethane, anthraquinonic and indigoid dyes can be applied (Abadulla et al. 2000). However, some substrate specificity can be found in laccase reactions, which limits the number of azo dyes that can be degraded. To solve this problem laccase/mediator systems are normally used to increase the decolorization rates (Bourbonnais et al. 1997, Rodríguez Couto et al. 2005, Camarero et al. 2005).

2. Decolorization of black liquor

The paper and pulp industry releases large volumes of colored black-liquors that contain toxic chlorinated lignin-degradation products. These products include chlorolignins, chlorophenols and chloroaliphatics (Ali and Sreekrishnan 2001). The mill effluents are highly alkaline and alter the pH of the soil and water bodies into which they are discharged.

LAP LAMBERT ACADEMIC PUBLISHING AG & CO. KG, DUDWELLER LANDSTR, GERMANY

The peroxidases play an important role in decolorization of black liquor. The mutant of Phanerochaete chrysosporium decolorize the bleach plant effluent from paper and pulp mills that lacked the ability to produce the peroxidases however MnP exhibited about 80% of the decolorizing activity (Frederick et al. 1991). Earlier studies demonstrated that marine fungi Sordaria fimicola (NIOCC #298) and Halosarpheia ratnagiriensis (NIOCC #321), which produced MnP and laccase, brought about 65-75 % decolorization of bleach plant effluent within 8 days (Raghukumar et al. 1996).

3. Decolorization of molasses spent wash

Molasses spent wash (MSW) is a by-product of sugar mills and alcohol distilleries, where the starting material is molasses. The colour of recalcitrant compounds is mostly dark brown, collectively termed as melanoidin, which are formed by the Maillard amino-carbonyl reaction (Wedzicha and Kaputo 1992). These compounds are toxic to many microorganisms including those generally involved in wastewater treatment processes (Kitts et al. 1993). The white-rot fungi Phanerochaete chrysosporium MnP-dependent was used to remove the color from MSW (Dehorter and Blondeau 1993) and laccase-dependent in Trametes versicolor (González et al. 2007). Furthermore the fungi are also capable of decolorizing MSW in the presence of seawater (D'Souza et al. 2006).

4. Decolorization of synthetic dyes and textile effluents

Textile industries effluents are detrimental on the aquatic life. Exhaustive reviews on decolorization of synthetic dyes (Wong and Yu 1999) and white-rot fungi and their lignin-degrading enzymes are used in the wastewaters and dye decolorizer (Fu and Viraraghavan 2001; Wesenberg et al. 2003). The role of laccase in terrestrial fungi in decolorization of dyes and dye wastewaters is undisputed (Wong and Yu 1999; Fu and Viraraghavan 2001; Wesenberg et al. 2003).

5. Treatment of effluents with crude/purified enzyme and immobilized enzymes

Lignin-degrading enzymes or H_2O_2 generating mechanism for degradation of lignin and decolorization of several effluents are widely reported (Wesenberg et al. 2003; Svabodová et al. 2008). The white-rot fungi showed an increased degradation of lignin and decolorization applications with partially purified or crude laccase (Wong and Yu 1999; Rodriguez et al. 1999). Immobilized lignin degrading enzymes have been extensively used for decolorization of textile effluents and synthetic dyes (Abadulla et al. 2000). Various supports such as alumina particles, chemically modified silica, amberlite and glass-ceramic have been used for this purpose (Abadulla et al. 2000). It has been demonstrated that treatment of azo dyes with free lignin degrading enzyme results in darkening of the solution due to coupling of the degraded products with the unreacted dyes (Zille et al. 2005).

6. Detoxification of effluents

Industrial effluents from paper and pulp mills and textile dye waste waters are toxic and mutagenic (Reddy 1995). Toxicity of several textile dyes, including azo compounds, was reduced by treatment with lignin degrading enzymes from Trametes hirsute (Abadulla et al. 2000). Eight white-rot fungi grown in green olives reduced phenolic content by nearly 70-75 % but phytotoxicity was not reduced (Aggelis et al. 2002). All of these fungi produced laccase and some of them produced MnP and LiP, however Rhizomucor pusillus strain RM7, a mucoralean fungus and a white-rot fungus Coriolus versicolor were shown to detoxify bleach plant effluent (Driessel and Christov 2001).

Bioremediation

Lignin degrading enzymes are useful for the removal of toxic compounds through oxidative enzymatic coupling of the contaminants, leading to insoluble complex structures (Wang et al 2002). Phenolic compounds are present in wastes from several industrial processes, such as coal conversion, petroleum refining, production of organic chemicals

LAP LAMBERT ACADEMIC PUBLISHING AG & CO. KG, DUDWELLER LANDSTR, GERMANY

and olive oil production (Aggelis et al 2003). Immobilized laccase was found to be useful to remove phenolic and chlorinated phenolic pollutants (Ehlers G.A. and Rose, P.D. 2005), due to the broad substrate range of the enzyme . The white rot fungi have been also perform oxidation of alkenes, carbazole, N-ethylcarbazole, fluorene, and dibenzothiophene in the presence of HBT and ABTS as mediators. Phenols contained in the OMW (olive-mill wastewater) have a structure similar to lignin, which makes them difficult to biodegrade. In a recent study, the treatment of OMW with several laccase-producing fungi led to the removal of up to 78% of the initial phenolic compounds in 12/15 days (Tsioulpas et al., 2002). This was associated to a decolorisation of the OMW from black to yellow-brown (Kissi et al, 2001) and to a decrease of the phytotoxicity, as described by the Germination Index parameter (Fountoulakis et al., 2002).

CONCLUSION

Microbial decolorization and degradation of colored effluents is a cost-effective and promising green technology for treatment of industrial effluents. Such effluents are mostly alkaline and rich in carbonates, chlorides and sulfates. The lignin degrading microbes and their enzymes have great potential for the application of remediation and dye decolorization of colored effluents.

Moreover, the lignin degrading microbes have been reviewed extensively for their bioremediation potentials. Synthetic dyes can be easily degraded in economic, effective, environmental and bio friendly manner. Moreover bioremediation of xenobiotics including synthetic dyes by various lignin degrading microbes will hopefully prove a green solution to the problem of environmental soil and water pollution in future. The isolation, purification and large-scale production of lignin-degrading enzymes with high activity or strain improvement to get maximum yield of such enzymes is the need of future for better environment.

REFERENCE

1. Abadulla, E., Tzanov, T., Costa, S., Robra, K-H., Cavaco-Paula, A. and Gübitz, G.M. (2000). Decolorization and detoxification of textile dyes with a laccase from Trametes hirsuta. Appl. Environ. Microbiol. 66, 3357-3362.

2. Aggelis G, Iconomou D, Christou M, Bokas D, Kotzailias S, Christou G, Tsagou V, Papanikolaou S (2003). Phenolic removal in a model olive oil mill wastewater using Pleurotus ostreatus in bioreactor cultures and biological evaluation of the process. Water Res. 37:3897-3904.

3. Aggelis, G., Ehaliotis, C., Nerud, F., Stoychev, I., Lyberotes, G., and Zervakis, G.I. (2002). Evaluation of white-rot fungi for detoxification and decolorization of effluents from the green olive debittering process. Appl. Microbiol. Biotechnol. 59, 353-360.

4. Akhtar M, Blanchette RA, Kirk TK (1997). In: Advances in Biochemical Engineering/Biotechnology (Eriksson KEL ed), Springer-Verlag, Germany, pp.159-195.

5. Alexandre, G; Zhulin, IB. Laccases are widespread in bacteria. Trends Biotechnol, 2000, 18, 41-42.

6. Ali, M. and Sreekrishnan, T.R. (2001). Aquatic toxicity from pulp and paper mill effluents: A review. Adv. Environ. Res. 5, 175–196.

7. Barreca, AM; Sjögren, B; Fabbrini, M; Galli, C; Gentili, P. Catalytic efficiency of some mediators in laccase-catalyzed alcohol oxidation. Biocat. Biotransform., 2004, 22, 105-112.

8. Berrocal, M; Ball, AS; Huerta, S; Barrasa, JM; Hernández, M; Pérez-Leblic, MI; Arias, ME. Biological upgrading of wheat straw through solid-state fermentation with Streptomyces cyaneus. Appl. Microbiol. Biotechnol., 2000, 54, 764-771.

9. Blanchette, RA. Degradation of lignocellulose complex in wood. Can J Bot, 1995, 73, S999-S1010.

10. Blanchette, RA; Cease, KR; Abad, AR. An evaluation of different forms of deterioration found in archaeological wood. Int. Biodeter Biodegrad, 1991, 28, 3-22.

11. Boudet, A. M., S. Kajita, et al. (2003). "Lignins and lignocellulosics: a better control of synthesis for new and improved uses." Trends Plant Sci. 8(12): 576-81.

12. Bourbonnais, R; Paice, MG; Freiermuth, B; Bodie, E; Borneman, S. Reactivities of various mediators and laccases with kraft pulp and lignin model compounds. Appl. Environ. Microbiol. 1997, 63, 4627-4632.

13. Brunow, G. Methods to reveal the structure of lignin. In: Hofrichter M, Steinbüchel A, editors. Biopolymers: lignin, humic substances and coal. Germany: Wiley-VCH; 2001; 89-116.

14. Camarero S, Ibarra D, Martinez M J, Martinez A T (2005). Lignin-derived compounds as efficient laccase mediators for decolorization of different types of recalcitrant dyes. Appl. Environ. Microbiol. 71:1775-1784.

15. Claus, H. Laccases and their occurrence in prokaryotes. Arch. Microbiol., 2003, 179, 145-150.

16. D'Souza, D.T., Tiwari, R., Sah, A.K. and Raghukumar, C. (2006). Enhanced production of laccase by a marine fungus during treatment of colored effluents and synthetic dyes. Enzyme Microb. Technol. 38, 504-511.

17. Daniel, G; Nilsson, T. Developments in the study of soft rot and bacterial decay. In: Bruce A, Palfreyman JW, editors. Forest products biotechnology. Great Britain: Taylor and Francis; 1998; 37-62.

18. Dehorter, B. and Blondeau, R. (1993). Isolation of an extracellular Mn-dependent enzyme mineralizing melanoidins from the white-rot fungus Trametes versicolor. FEMS Microbiol. Lett. 109, 117-122.

19. Dix NJ, Webster J (1995). Fungal Ecology, Chapman & Hall, Cambridge, Great Britain.

20. Driessel, B.V. and Christov, L. (2001). Decolorization of bleach plant effluent by mucoralean and white-rot fungi in a rotating biological contactor reactor. J. Biosci. Bioeng. 92, 271-276.

21. Eaton, RA; Hale, MDC. (1993). Wood: decay, pests and protection. Cambridge, Great Britain: Chapman and Hall.

22. Eggert, C; Temp, U; Dean, JFD; Eriksson, KEL. (1996). Laccase-mediated formation of the phenoxazinone derivative, cinnabarinic acid. FEBS Lett., 376, 202-206.

23. Ehlers G A, Rose P D (2005). Immobilized white-rot fungal biodegradation of phenol and chlorinated phenol in trickling packed-bed reactors by employing sequencing batch operation. Bioresource Technology. 96:1264-1275.

24. Falcón, MA; Rodríguez, A; Carnicero, A; Regalado, V; Perestelo, F; Milstein, O; de la Fuente, G. (1995). Isolation of microorganisms with lignin transformation potential from soil of Tenerife island. Soil Biol. Biochem, 27, 121-126.

25. Foster, C. E., T. M. Martin, et al. (2010). "Comprehensive compositional analysis of plant cell walls (Lignocellulosic biomass) part I: lignin." J. Vis. Exp. (37).

26. Fountoulakis, MS., SN. Dokianakis, ME. Kornaros, GG. Aggelis, and Lyberatos. (2002). Removal of phenolics in olive mill wastewaters using the white-rot fungus Pleurotus ostreatus. Water Res. 36:4735-4744.

27. Frederick, Jr.C. M., Dass, S.B., Grulke, E.A. and Reddy, C.A. (1991). Role of manganese peroxidases (MnP) and lignin peroxidases (LiP) of Phanerochaete chrysosporium in the decolorization of kraft bleach plant effluent. Appl. Environ. Microbiol. 57, 2368-2375.

28. Fu, Y. and Viraraghavan, T. (2001). Fungal Decolorization of dye wastewaters: a review. Bioresource Technol. 79, 251-262.

29. Galli, C; Gentili, P. Chemical messengers: mediated oxidations with the enzyme laccase. J. Phy. Org. Chem., 2004, 17, 973-977.

30. Gold, MH; Alic, M. Molecular biology of the lignin-degrading basidiomycete Phanerochacte chrysosporium. Microbiol. Rev., 1993, 57, 605-22.

31. Gonzáles, T., Terrón, M.C., Yagüe, S., Junca, H., Carbajo, J.M., Zapico, E.J., Silva, R., Arana- Cuenca, A., Téllez, A. and Gonzáles, A.E. (2008). Melanoidin-containing wastewaters induce selective lacasse gene expression in the white-rot fungus Trametes sp. I-62. Res. Microbiol. 159, 103-109.

32. Hammel, KE. Fungal degradation of lignin. In: Cadisch, G and Giller, KE, editors. Driven by Nature: Plant Litter Quality and Decomposition. Wallingford, UK: CAB International; 1997; 33–45.

33. Hatakka A (2001). In: Biopolymers (Hofrichter M, Steinbüchel A eds), Wiley-VCH, Germany, pp. 129-180.

34. Hatakka, A. (1994). Lignin-modifying enzymes from selected white-rot fungi: production and role in lignin degradation. FEMS Microbiology Reviews, , 13, 125–135.

35. Hoegger, PJ; Kilaru, S; James, TY; Thacker, JR; Kuees, U. Phylogenetic comparison and classification of laccase and related multicopper oxidase protein sequences. FEBS J., 2006, 273, 2308-2326.

36. Hofrichter, M. (2002). Review: lignin conversion by manganese peroxidase (MnP). Enzyme Microbial. Technol., 30, 454-466.

37. Karkonen, A. and S. Koutaniemi (2010). Lignin biosynthesis studies in plant tissue cultures. J. Integr Plant Biol. 52(2), 176-85.

38. Kirk, TK; Popp, JL; Kalyanaraman, B. (1990). Lignin peroxidase oxidation of Mn^{2+} in the presence of veratryl alcohol, malonic or oxalic acid and oxygen. Biochem. J, 29, 10475-10480.

39. Kirk, TK; Popp, JL; Kalyanaraman, B. Lignin peroxidase oxidation of Mn^{2+} in the presence of veratryl alcohol, malonic or oxalic acid and oxygen. Biochem. J., 1990, 29, 10475-10480.

40. Kissi, M., M. Mountadar, O. Assobhei, E. Gargiulo, G. Palmieri, P. Giardina, and G. Sannia. (2001). Roles of two white-rot basidiomycete fungi in decolorisation and detoxification of olive mill waste water. Appl. Microbiol. Biotechnol. 57:221-226.

41. Kitts, D.D., Wu, C.H., Stich, H.F. and Powrie, W.D. (1993). Effect of glucose-glycine Maillard reaction products on bacterial and mammalian cells mutagenesis. J. Agri. Food Chem. 41, 293-301.

42. Lacey, J. Actinomycetes as biodeteriogens and pollutant of the environment. In: Goodfellow M, Williams ST, Mordarski M, editors. Actinomycetes in biotechnology. Great Britain: Acad Press; 1988; 359-432.

43. Mercer, DK; Iqbal, M; Miller, PGG; McCarthy, AJ. Screening actinomycetes for extracellular peroxidase activity. Appl. Environ. Microbiol. 1996, 62, 2186-2190.

44. Orth AB, Pease EA, Tien M (1994). In: Biological Degradation and Bioremediation of Toxic Chemicals (Chaudhry GR ed), Chapman & Hall, London, pp. 345-363.

45. Paszczynski, A; Crawford, RL. (1995). Potential for bioremediation of xenobiotic compounds by the white-rot fungus Phanerochaete chrysosporium. Biotechnol. Pro., 11, 368-379.

46. Pease, EA; Tien, M. Heterogeneity and regulation of manganese peroxidases from Phanerochaete chrysosporium. 1992, 174, 3532-3540.

47. Perestelo, F; Rodriguez, A; Pérez, R; Carnicero, A; de la Fuente, G; Falcón, MA. (1996). Isolation of a bacterium capable of limited degradation of industrial and labelled, natural and synthetic lignins. World J. Microbiol. Biotechnol., 12, 111-112.

48. Pointing, SB. Feasibility of bioremediation by white-rot fungi. Appl. Microbiol. Biotechnol., 2001, 57, 20-33.

49. Raghukumar, C., Chandramohan, D., Michel, F.C. Jr. and Reddy, C.A.(1996). Degradation of lignin and decolorization of paper mill bleach plant effluent (BPE) by marine fungi. Biotechnol. Lett.18, 105-108.

50. Reddy, C.A. (1995). The potential of white-rot fungi in the treatment of pollutants. Curr. Opin. Biotechnol. 6, 320-328.

51. Rodríguez Couto S, Sanromán M, Gübitz G M (2005). Influence of redox mediators and metal ions on synthetic acid dye decolourization by crude laccase from Trametes hirsuta. Chemosphere. 58:417-422.

52. Rüttimann, C; Vicuña, R; Mozuch, MD; Kirk, TK. Limited bacterial mineralization of fungal degradation intermediates from synthetic lignin. Appl. Environ. Microbiol., 1991, 57, 3652-3655.

53. Schlosser D, Hofer C (2002). Laccase-catalyzed oxidation of Mn2+ in the presence of natural Mn3+ chelators as a novel source of extracellular H2O2 production and its impact on manganese peroxidase. Appl Environ Microbiol 68:3514-3521.

54. Scott GM, Akhtar M (2001). In: Lignin, Humic Substances and Coal (Hofrichter M, Steinbüchel A eds), Wiley-VCH, Germany, pp. 181-207.

55. Sjöström, E. (1993). Wood chemistry, fundamentals and applications, New York / London: Acad Press.

56. Spiker, JK; Crawford, DL; Thiel, EC. Oxidation of phenolic and non-phenolic substrates by the lignin peroxidase of Streptomycete viridosporus T7A. Appl. Microbiol. Biotechnol., 1992, 37, 518-523.

57. Sundaramoorthy, M; Kishi, K; Gold, MH; Poulas, TL. The crystal structure of manganese peroxidase from Phanerochaete chrysosporium 2.06Å resolution. J. Biol. Chem., 1994, 269, 32759-32767.

58. Susana, RC; José, LTH. Industrial and biotechnological applications of laccases: a review. Biotechnol. Adv., 2006, 24, 500-513.

59. Suzuki, T; Endo, K; Iro, M; Tsujibo, H; Miyamoto, K; Inamori, Y. A thermostable laccase from Streptomyces lavendulae REN-7: purification, characterization, nucleotide sequence, and expression. Biosci. Biotechnol. Biochem., 2003, 67, 2167-2175.

60. Svobodová, K., Majcherczyk, A., Novotný, C. and Kües, U. (2008). Implication of myceliumassociated laccase from Irpex lacteus in the decolorization of synthetic dyes. Bioresour. Technol. 99, 463-471.

61. Thurston, CF. The structure and function of fungal laccases. Microbiol., 1994, 140, 19-26.

62. Tsioulpas, A., D. Dimou, D. Iconomou, and G. Aggelis. (2002). Phenolic removal in olive oil mill wastewater by strains of Pleurotus spp. in respect to their phenol oxidase (laccase) activity. Bioresour Technol. 84:251-257.

63. Urzúa, U; Fernando, LL; Lobos, S; Larraín, J; Vicuña, R. Oxidation reactions catalyzed by manganese peroxidase isoenzymes from Ceriporiopsis subvermispora. FEBS Lett., 1995, 4, 132-136.

64. Vasdev, K; Kuhad, RC. (1994). Decolorization of poly R-478 (polyvinylamine sulphonate anthra pyridone) by Cyathus bulleri. Folia Microbiol, 39, 61-64.

65. Vasdev, K; Kuhad, RC; Saxena, RK. Decolorization of triphenylmethano dyes by the bird's nest fungus Cyathus bulleri. Curr. Microbiol., 1995, 30, 269-272.

66. Vicuña, R. Ligninolysis: a very peculiar microbial process. Mol. Biotechnol., 2000, 14, 173-176.

67. Vicuña, R; González, B; Seelenfreund, D; Rüttimann, C; Salas, L. Ability of natural bacterial isolates to metabolize high and low molecular weight lignin derived molecules. J. Biotechnol., 1993, 30, 9-13.

68. Vogt, T. (2010). Phenylpropanoid biosynthesis. Mol. Plant 3(1), 2-20.

69. Wang C J, Thiele S, Bollag J M (2002). Interaction of 2,4,6-Trinitrotoluene (TNT) and 4-Amino-2,6-Dinitrotoluene with Humic Monomers in the Presence of Oxidative Enzymes. Arch. Environ. Contam. Toxicol. 42:1- 8.

70. Wariishi, H; Dunford, HB; MacDonald, ID; Gold, MH. Manganese peroxidase from the lignin-degrading basidiomycetes Phanerochaete chrysosporium: transient-state kinetics and reaction mechanism. J. Biol. Chem., 1989, 264, 3335-3340.

71. Wedzicha, B.L. and Kaputo, M.T. (1992). Melanoidins from glucose and glycine: composition, characteristics and reactivity towards sulphite ion. Food Chemistry 43, 359-367.

72. Wesenberg, D., Kyriakides, I. and Agathos, S.N. (2003). White-rot fungi and their enzymes for the treatment of industrial dye effluents. Biotechnol. Adv. 22, 161-187.

73. Wong, KKY; Mansfield, SD. (1999). Enzymatic processing for pulp and paper manufacture-a review. Appita J, 52, 409-418.

74. Wong, Y. and Yu, J. (19990. Laccase catalyzed decolorization of synthetic dyes. Water Res. 33, 3512-3520.

75. Zille, A., Górnacka, B., Rehorek, A. and Cavaco-Paulo, A. (2005). Degradation of azo dyes by Trametes villosa laccase over long peroids of oxidative conditions. Appl. Environ. Microbiol. 71, 6711-6718.

76. Zimmermann, W. (1990). Degradation of lignin by bacteria. J. Biotechnol., 13, 119-130.

CHAPTER-6

Agrobacterium tumefaciens - Natural source for plant transformation

Ashish Dhyani*, Rishi. K. Saxena#, Vijendra Mishra* and Pallavi Sharma#

#Department of Microbiology, Bundelkhand University, Jhansi (U.P) 284128 India.

*Department of Dairy Microbiology, Anand University, Gujarat

ABSTRACT

Agrobacterium tumefaciens is widespread naturally occurring soil bacteria that causes crown gall, and has ability to introduce new genetic material into the plant cell. Almost three decades before, the molecular transformation by this bacterium as a vector to create transgenic plants was considered as a beautiful wish, but because of the advancement of microbial biotechnology, this wish is now being routinely utilised by many plant molecular laboratories. Members of the genus Agrobacterium are ubiquitous in nature as a soil microflora, the vast majority of which are saprophytic, surviving primarily on decaying organic matters. Thus, the use of Agrobacterium to genetically transform plants has advanced from dream to a reality. Modern agricultural biotechnology is heavily dependent on using Agrobacterium to create transgenic plants, and it is difficult to think of an area of plant science research that has not benefited from this technology. However, there remain many challenges. Many economically important plant species, or elite varieties of particular species, remain highly recalcitrant to Agrobacterium-mediated transformation. This chapter will put insight on A.tumefaciens mediated transformation.

Keywords : Agrobacterium tumefaciens, LPS, Vir genes, plant transformation,

INDEX

1. AGROBACTERIUM TUMEFACIENS : AN INTRODUCTION
2. T-DNA INTEGRATION AND EXPRESSION FOR CROWN GALL FORMATION
3. BROAD HOST RANGE AND BACTERIAL COLONIZATION

4. GENOME STRUCTURE AND GENES IN THE T-DNA

5. AGROBACTERIUM MEDIATED PLANT TRANSFORMATION PROCESS IN LAB

6. ECONOMICAL   AND   AGRICULTURAL   IMPORTANCE   OF A.TUMEFACIENS

7. TREATMENT AND CONTROL MECHANISM

Introduction

Agrobacterium tumefaciens is the most studied bacteria in this genus, which is a part of microflora and able to transform new genetic material into the plant cell[1] . Agrobacterium is a Gram negative, non- sporeforming, rod shaped, flagellated bacteria and is an important tool in genetic engineering for the introduction of foreign genes into plant tissue. The tumour-inducing genes are usually replaced with the gene of interest, and a marker gene (e.g. the antibiotic resistance gene, herbicide resistance gene) is added to enable selection of transformed cells. A.tumefaciens has the exceptionally ability to transfer a particular DNA segment (T-DNA) of the tumor-inducing (Ti)[2] plasmid into the nucleus of the infected cells where it is then stably integrated into the host genome and transcribed leading the plant towards disease called Crown gall[3-4] . Major infection and existence of its prevalence are found in dicotyledonous plants by Agrobacterium tumefaciens [5-8.]

Agrobacterium tumefaciens can generally be found on and around root surfaces known as the rhizosphere. There it seems to use nutrients that leak from the root tissue. It will infect the tissue at wound sites formed from transplanting seedlings, burrowing animals or bugs, etc, Thus when a wound opens on the plant tissue, the motile cells of bacterium swims towards concentration of phenolic compounds and also small metabolites such as glucose and amino acids which are usually exuded from plant wounds. VirA is the receptor which senses the presence of these substances and guides the flagella towards the source of these substances. Activated VirA then phosphorylates VirG, which is responsible for preparing the cell for infecting host plant tissue, including the chemotactic response of swimming towards up the gradient of phenolics to find its host.bacteria contain two types of genes: the oncogenic genes, responsible for encoding the enzymes involved in the synthesis of auxins and cytokines and cause tumours, other genes encodes for the synthesis of opines. These compounds, are synthesized and excreted by the crown gall cells and consumed by bacteria as a carbon and nitrogen source[9]-[10].

Agrobacterium strains use different carbohydrates and are classified into three main biovars. The differences among biovars are mainly determined by the genes on the circular chromosome. In order to be virulent Agrobacterium strains must harbour a tumour inducing plasmid (Ti). This plasmid is introduced into the host plant genome for the induction and maintenance of tumours by the virulent strains[11]-[14].Lippincott et al have shown that the primary event in crown gall tumour induction is the attachment of the bacterial cells to the host plant cell wall[15]-[16]. This bacterial adherence has been confirmed by various groups of workers using different methods including electron microscopy[17]-[21]. Studies on A.tumefaciens led the revolution in agriculture biotechnology and made easiest way to transformed a genetic element to make a new transgenic plant to cope against the pathogenic affect of harmful microflora in rhizosphere or introduced from external environment.

T-DNA integration and expression for crown gall formation:

The A.tumefaciens genome has a very unusual structures, four DNA elements carries around 5400 genes- a circular chromosome, a linear chromosome and two smaller circular structures called Plasmids. Many bacteria have circular chromosomes and some have linear chromosomes, but Agrobacteria are the only known species to have both structures together.

There are series of sequential actions involved in the pathogenicity or the transformation of the plant response (fig 1), beginning with the chemotaxis followed by gene responses and behaviour of the plant.

Plant secretes sugars and other components in their roots, which attracts many microorganism in the rhizosphere including A.tumefaciens for their nutrient supplements, this process of attraction is called chemotaxis.A. tumefaciens can use a variety of substrates for energy and carbon, but it is especially evolved to use a class of chemicals called opines, which are amino acid-like compounds that are intermediates of metabolism in most organisms. A. tumefaciens forces the plants it infects to produce opines, a molecule the bacteria use as an energy and carbon source. There are many types of opines which it can use, such as nopaline, agropine, mannopine (which are common), and chrysopine, deoxy-fructosyl-oxo-proline (which are uncommon). It is believed that each strain of A. tumefaciens can only metabolize one type of opine, and contains genes for its synthesis (usually in its T-DNA which it transfers to its plant host) and catabolism, although this is not strictly true.

The infection of Agrobacterium to the host initates with the involvement of a set of chromosome genes (chv) those involved in attachment of bacteria to plant cells and Ti plasmid-encoded vir genes, helping in the generation, transfer, and integration of T stands into the plant genome [22].

The Vir genes consisting of seven genetically identified operons, VirA,-B,-C,-D,-E,-G and H[23-24]., are responsible for most steps in the transfer of T-DNA from the bacterium into the plant cell. These genes are only expressed in the presence of plant exudates containing low-molecular weight phenolic compounds such as acetosyringone[25]

The virA and virG gene products are required for the regulation of vir regulon on the tumor-inducing (Ti) plasmid of A.tumefaciens. As stated above, VirA is a membrane bound protein which is homologous to the sensor molecules of the other two component regulatory systems, this signal passes down and gets autophosphorylation of the receptor begins with the phosphorylation at His-474 residue[26]. The signal cascade starts intracellularly, and inturn activates the cytoplasmic protein VirG by phosphorylating it at Asp-52 position[27]. Now, VirG which is a DNA binding protein[28-29] and act as a transcriptional activator of vir operons[30-32].

LAP LAMBERT ACADEMIC PUBLISHING AG & CO. KG, DUDWELLER LANDSTR, GERMANY

Thus, as per the action mediated by VirA and VirG are also known as response regulator[33]. Thus VirA and VirG sense phenolic compounds from wounded plant cells and induce expression of other virulence genes, VirD1 and VirD2 that excise the T-DNA from its adjacent sequences. Subsequent to transfer to the plant cell, the virulence protein VirD2, through its nuclear localization signal (NLS), is believed to guide the T-DNA to the nucleus[34].

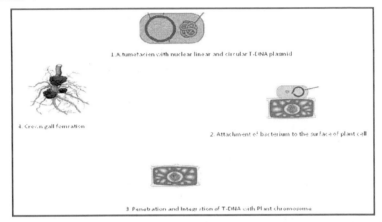

Fig: 6.1- Sequential progressive formation of crown gall disease in infected plant with A.tumefaciens, leading the T-DNA migration and integration due to genes responses, cause the uncontrolled proliferation of the plant cells.

The VirD1 protein, by itself localized in the cytoplasm, moved to the nucleus when coexpressed with the VirD2 protein, suggesting VirD1–VirD2 interaction. VirC is responsible to make a binding to the overdrive region to promote high efficiency T-strand synthesis. VirE2 helps T-strand from nuclease attack, and intercalates with lipids to form channels in the plant membranes through which the T-complex passes. While VirE1 acts as a chaperone which stabilises VirE2 in the Agrobacterium. VirB and Vir D4 assembles into secretion system which spans the inner and outer bacterial membranes. Required for export of the T-complex and VirE2 into the plant cell. The cluster of genes function in Ti plasmid made plant to transformed to the tumurous growth or uncontrolled cellular proliferation. Thus genes in the Ti plasmid are tumerogenic in nature for plants mainly dicotyledonous plants.

Broad host range and Bacterial Colonization

A.tumefaciens has perhaps the largest host range among any other plant pathogenic bacterium, causing crown gall disease. Many cultivated monocots and legumes are not a good source for A.tumefaciens habitat, Although some of these recalcitrant plants (eg Rice) can be transformed by Agrobacterium vectors under controlled laboratory conditions[35]. Agrobacterium genus has many biological important species including A.radiobacter is an "avirulent" species, A.tumefaciens causes crown gall disease, A.rhizogenes causes hairy root disease, A.vitis causes galls on grape, and A.rubi causes cane gall disease[36]. The infection caused by A.tumefaciens in its host plants range from plum[37], peach[38], grapes[39], aspen[40], rose[41], peas[42], cucurbits[43], soybeans[44]-[45], and others. Entering of A.tumefaciens in its various host range detoxify hydrogen peroxide by its enzyme KatA, which converts hydrogen peroxide into oxygen and water, thus affect host plant defence mechanism. Hydrogen peroxide work as a primary component of the plant defence mechanism that has both direct germicidal activity and a signalling function [46]. The resistant or control mechanism at the beginning of penetration or adhesion may prevent the lethal cause of its effect to the plant, as many mutagenesis studies shown that non-attaching mutants loss the Tumor-inducing capacity[47]-[50]. Research evidences proven the fact of the role of lipopolysaccharides of A.tumefaciens cell surface plays a significant role in the development of colonizing process. The LPS are the integral part of the outer membrane and include lipid A membrane anchor and the O-antigen polysaccharide in their composition. A.tumefaciens lacking lipid anchor and having strong anionic nature and tight association with the cell. There are some evidences indicating that capsular polysaccharides may play a specific role during the interaction with the host plant.

LAP LAMBERT ACADEMIC PUBLISHING AG & CO. KG, DUDWELLER LANDSTR, GERMANY

Genome Structure and Genes in the T-DNA

A.tumefaciens shows the unusual genetic composition containing with - a circular chromosome, a linear chromosome and two smaller circular structures called Plasmids. Its genome has a total of 5.7 million base pairs, with 2.8 million residing on its circular chromosome and 2.1 million residing on its linear chromosome[51]. The main genes those are responsible for A.tumefaciens survival are encodes from circular chromosomes (see fig2). although through evolution some essential genes have migrated to the linear chromosome. Based on sequence analysis, it was determined that the linear chromosome was derived from a plasmid that was transformed into the bacteria a long time ago (Goodner, et al).

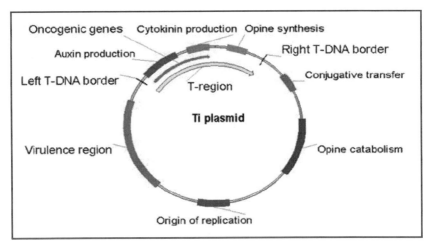

Fig: 6.2- A circular chromosome indicating its various sites for multigene association.

In addition to the two chromosomes, strain C58 also contains two plasmids, pTiC58 (generically called Ti) and pAtC58 (also called the "cryptic plasmid"). pTiC58 contains genes necessary for its pathogenicity against plants (Goodner, B. et al.), including the T-

LAP LAMBERT ACADEMIC PUBLISHING AG & CO. KG, DUDWELLER LANDSTR, GERMANY

DNA which is injected into the plant and causes it to produce opines, along with accessory proteins which helps the T-DNA enter and transform the plant cell into a tumor cell. It is believed that pAtC58 contains genes essential for opine catabolism[52] or its ability to use opines as an energy source, which is important for its lifestyle as a pathogen.

In order to cause gall formation, the T-DNA encodes genes for the production of auxin or indole-3-acetic acid via the IAM pathway. This biosynthetic pathway is not used in many plants for the production of auxin, so it means the plant has no molecular means of regulating it and auxin will be produced constitutively. Genes for the production of cytokinins are also expressed. This stimulates cell proliferation and gall formation.

The T-DNA contains genes for encoding enzymes that cause the plant to create specialized amino acids which the bacteria can metabolize, called opines. Opines are a class of chemicals that serve as a source of nitrogen for A. tumefaciens, but not for most other organisms. The specific type of opine produced by A. tumefaciens C58 infected plants is nopaline.

Agrobacterium mediated plant transformation process in Lab

The Agrobacterium-mediated plant transformation involves a number of steps: Below representation shows how plant can be transformed in the various steps for the choosen requirement.

Isolation of the gene of interest from the source organism
↓
Insertion of the transgene into the Ti Plasmid
↓
Introduction of the T-DNA containing plasmid into Agrobacterium
↓
Mixture of the transformed Agrobacterium with plant cells to allow transfer of T-DNA into plant chromosome
↓
Regeneration of the transformed cells into genetically modified (GM) plants
↓
Testing for traits performance or transgene expression at lab, greenhouse and field level.

Economical and Agricultural Importance of A.tumefaciens

The hassle free gene delivery system of A.tumefaciens at the T region of its genomic composition makes a significant tool for the agricultural biotechnology, which lead a revolution in last two decades, the T-complex is able to carry any gene of interest and can integrate with the targeted plant genome with a high degree of success. The reason for this is because unlike other mobile genetic elements such as transposons and retroviruses, the T-DNA strand does not encode functions required for movement and integration of the DNA. Therefore the T-DNA strand can be replaced by a gene of interest which will be inserted automatically into the host plant nucleus with a high degree of success and with little human intervention. This process is usually much more efficient than traditional methods of genetic modification. Thus, it has proven and successfully mediates the foreign DNA into plant genomes for genetic modification. This use of Agrobacterium is based on its unique capacity for "trans-kingdom sex", i.e. transfer of genetic material between prokaryotic and eukaryotic cells like from yeast[53] to mushrooms and filamentous fungi to phytopathogenic fungi to human cells. Several different plant species have already been successfully transformed, including Lettuce , Rice and Tomato.The plasmid T-DNA that is transferred to the plant is an ideal vehicle for genetic engineering. This is done by cloning a desired gene sequence into the T-DNA that will be inserted into the host DNA. This process has been performed using firefly luciferase gene to produce glowing plants. This luminescence has been a useful device in the study of plant chloroplast function and as a reporter gene. It is also possible to transform Arabidopsis thaliana by dipping their flowers into a broth of Agrobacterium: the seed produced will be transgenic. Under laboratory conditions the T-DNA has also been transferred to human cells, demonstrating the diversity of insertion application. The mechanism by which Agrobacterium inserts materials into the host cell by a type IV secretion system is very similar to mechanisms used by pathogens to insert materials (usually proteins) into human cells by type III secretion. It also employs a type of signaling conserved in many Gram-negative bacteria called quorum sensing. This makes Agrobacterium an important topic of medical research as well.

LAP LAMBERT ACADEMIC PUBLISHING AG & CO. KG, DUDWELLER LANDSTR, GERMANY

Treatment and control Mechanism

Once A.tumefaciens infects a plant, the bacterium travels throughout the root system, and can wipe out an entire crop. The only option as to destroy the plants. The practice of replacing the tumor-inducing genes with other DNA began in the 1970s and led to the widespread use of bacterium in research. Agrobacterium elicits neoplastic growths (called crown gall tumors) that affect most dicotyledonous plants. Different control measures have been used against the infection of A.tumefaciens called crown gall. Few most prominent examples of the control of crown gall disease includes treatment of virulent A.tumefaciens with the avirulent strain of A.radiobacter K84 strain[54], soil fumigation[55], chemical[56], Antibiotics[57] and soil solarisation[58-59]. Among all these controlling technique, soil solarisation was found to most effective and safer for plant. While natural A.tumefaciens populations that resist K84 treatment are known to exist, and K84 is not effective on infected asymptomatic plants[60-61].Furthermore, Soil fumigation gives incomplete control of crown gall and was reported to induce an unexpected increase of disease incidence[62]. In addition to the use of chemicals and antibiotics shows incomplete crown gall control and are often phytotoxic[63].

The primary control for development of crown gall is to prevent plant from any external injury. Carful cultural practices that prevent unnecessary plant wounding can significantly reduce crown gall by denying A.tumefaciens an opportunity to introduce T-DNA into plant cells[64]. External wound approaches from subfreezing winter temperature, insects and nematodes can be crucial in preventing natural wounds that can act as site of infection[65-66]. it is apparent that a suite of host plant genes is required for efficient Agrobacterium T-DNA transfer and integration[67-68]. Specific cell wall proteins (e.g. vitronectin-like proteins) are probably required for bacterial attachment, nuclear import machinery (e.g. importin-a andVIP1) is required for T-DNA subcellular trafficking and components of DNA repair and recombination pathways (potentially including histone H2A-1) appear to be required for T-DNA integration. Posttranscriptional gene silencing (PTGS) of any of these plant genes could generate disease resistance by blocking the process of

Agrobacterium transformation[69-70]. Indeed, individual transgenic Arabidopsis plants expressing VIP1 antisense RNA, importin a1 antisense RNA or histone H2A-1 antisense RNA display crown gall disease resistance .

References

1. Agrios, G.N. (1997) Plant Pathology, 4th edn, Academic Press.

2. Alcornero R., 1980. Crown gall of peaches from Maryland, South Carolina, and Tennessee and problems with biological control. Plant Disease 64: 835-838.

3. Bailey, M. A., H. R. Boerma, and W. A. Parrott. 1994. Inheritance of Agrobacterium tumefaciens-induced tumorigenesis of soybean. Crop Sci.34:514–519.

4. Beneddra, T. et al. (1996) Correlation between susceptibility to crown gall and sensitivity to cytokinin in aspen cultivars. Phytopathology 86,225–231

5. Binns, A.N. and Thomashow, M.F. (1988). Cell biology of Agrobacterium infection and transformation of plants. Annual Review of Microbiology 42: 575-606.

6. Biserka Relic, Mirjana Andjelkovic, Luca Rossi Yoshikuni Nagamine and Barbara Hohn. Interaction of the DNA modifying proteins VirD1 and VirD2 of Agrobacterium tumefaciens: Analysis by subcellular localization in mammalian cells. Proc. Natl. Acad. Sci. USA,Vol. 95, pp. 9105–9110, August 1998

7. Bliss, F.A. et al. (1999) Crown gall resistance in accessions of Prunus species. HortScience 34, 326–330

8. Bradley L.R., Kim, J.S. and Matthysse, A.G. (1997). Attachment of Agrobacterium tumefaciens to Carrot Cells and Arabidopsis wound sites is correlated with the presence of a cell-associated, acidic polysaccharide. Journal of Bacteriology 179:5372-5379.

9. Burr, T.J. and Otten, L. (1999) Crown gall of grape: biology and disease management. Annu. Rev. Phytopahol. 37, 53–80.

LAP LAMBERT ACADEMIC PUBLISHING AG & CO. KG, DUDWELLER LANDSTR, GERMANY

10. Burr, T.J. et al. (1998) Biology of Agrobacterium vitis and the development of disease control strategies. Plant Dis. 82, 1288–1297

11. Canfield M.L., Pereira C., Moore L.W.,1992. Control of crown gall in apple (Malus) rootstocks using Copac E and Terramycin. Phytopathology 82: 1153.

12. Cangelosi, G.A., Hung, L., Puvanesarajah, V., Stacey, G.,Ozga. D.A., Leigh, J.A. and Nester, E.W. (1987). Common loci for Agrobacterium tumefaciens and Rhizobium meliloti exopolysaccharide synthesis and their role in plant interaction. Journal of Bacteriology 169:2086-2091.

13. Chilton, M. D., Drummond, H. J., Merlo, D. J., Sciaky, D., Montoya, A. L., Gordon, M. P. and Nester,E. W. (1977) Cell, 11, 263.

14. Das, A., S. Stachel, P. Ebert, P. Allenza, H. Montoya, and E. W. Nester. 1986.Promoters of Agrobacterium tumefaciens Ti plasmid virulence genes. Nucleic Acids Res. 14:1355–1364).

15. Deep I.W., McNeilan R.A., Macswan I.C., 1968. Soil fumigants tested for control of crown gall. Plant Disease Reporter 52: 102-105.

16. Douglas C.J., Halperin, W.and Nester, E.W. (1982).Agrobacterium tumefaciens mutants affected in attachment to plant cell. Journal of Bacteriology 152:1265-1275.

17. Gelvin B.S (2003). Agrobacterium-Mediated plant transformation: the Biology behind the "gene jockeying" tool. Microbiology and molecular biology reviews 67(1): 16-37.

18. Gelvin, S.B. (2000) Agrobacterium and plant genes involved in T-DNA transfer and integration. Annu. Rev. Plant Physiol. PlantMol. Biol. 51,223–256

19. Gheysen, G., G. Angenon, and M. Van Montagu. 1998. Agrobacterium-mediated plant transformation: a scientifically intriguing story with significant applications, p. 1–34. In L. K. Harwood (ed.), Transgenic plant research. Academic Publishers, Amsterdam, The Netherlands.).

20. Goodner.B et al, Genome Sequence of the Plant Pathogen and Biotechnology Agent Agrobacterium tumefaciens C58. Science 14 December 2001: Vol. 294 no. 5550 pp. 2323-2328,DOI: 10.1126/science.1066803

21. Grimm R., Sule S., 1981. Control of crown gall (Agrobacterium tumefaciens Smith & Townsend) in nurseries. In:Lozano JC, Gwin P, (eds). Proceedings of the 5th InternationalConference on Plant Pathogenic Bacteria, Cali,Colombia, 531-537.

22. Grimm R., Sule S., 1981. Control of crown gall (Agrobacterium tumefaciens Smith & Townsend) in nurseries. In:Lozano JC, Gwin P, (eds). Proceedings of the 5th InternationalConference on Plant Pathogenic Bacteria, Cali,Colombia, 531-537.

23. Haico van Attikum1, Paul Bundock1 and Paul J. J. Hooykaas . Non-homologous end-joining proteins are required for Agrobacterium T-DNA integration.The EMBO Journal (2001) 20, 6550 - 6558 doi:10.1093/emboj/20.22.6550

24. Hooykaas, P.J.J. and Shilperoort, R.A. (1992). Agrobacterium and plant genetic engineering. Plant Molecular Biology 19:15-38.

25. Huang, Y., P. Morcl, B. Powell, and C. I. Kado. 1990. Vir A, a coregulator of Ti-specified virulence genes, is phosphorylated in vitro. J. Bacteriol. 172:1142–1144.

26. Jin, S., R. K. Prusti, T. Roitsch, R. G. Ankenbauer, and E. W. Nester. 1990.Phosphorylation of the VirG protein of Agrobacterium tumefaciens by the autophosphorylated VirA protein: essential role in biological activity of VirG. J. Bacteriol. 172:4945–4950.

27. Jin, S., T. Roitsch, P. J. Christie, and E. W. Nester. 1990. The regulatory VirG protein specifically binds to a cis-acting regulatory sequence involved in transcriptional activation of Agrobacterium tumefaciens virulence genes. J.Bacteriol. 172:531–537).

28. Kanemoto, R. H., A. T. PoweUl, D. E. Akiyoshi, D. A. Regier,R. A. Kerstetter, E. W. Nester, M. C. Hawes, and M. P. Gordon.1989. Nucleotide sequence and analysis of the plant-inducible locus pinF from Agrobacterium tumefaciens. J. Bacteriol. 171:2506-2512.

LAP LAMBERT ACADEMIC PUBLISHING AG & CO. KG, DUDWELLER LANDSTR, GERMANY

29. Katan J., 1980. Solar pasteurization of soil for disease control:status and prospects. Plant Disease 64: 450-454.

30. Leroux, B., M. F. Yanofsky, S. C. Winans, J. E. Ward, S. F. Ziegler, and E. W. Nester. 1987. Characterization of the virA locus of Agrobacterium tumefaciens:a transcriptional regulator and host range determinant. EMBO J.6:849–856.

31. Lippincott, J. A. and Lippincott, B. B. (1977) Cell Wall Biochemistry Related to Specificity in Hostpathogen Interactions, eds B. Solheim and J. Raa, (Norway: Universitetsforlaget, Oslo) p. 439.

32. Lippincott, J. A. and Lippincott, B. B. (1980) Bacterial Adherence, ed. E. H. Beachey, (London: Chapman and Hall) p. 377.

33. Manigault, P. (1970) Ann. Microbiol. Inst. Past. Paris, 119, 347.

34. Matthysse, A. G. (1978) Infect. Immun., 22, 516.

35. Mauro, A. O., T. W. Pfeiffer, and G. B. Collins. 1995. Inheritance of soybean susceptibility to Agrobacterium tumefaciens and its relationship to transformation.Crop Sci. 35:1152–1156.

36. Nadolska-Orczyk, A. et al. (2000) Agrobacterium-mediated transformation of cereals: from technique development to its applications. Acta Physiol. Plant. 22, 77–78

37. Nester, E.W., Gordon, M.P., Amasino, R.M. and Yanofsky, M.F. 1984). Crown gall: a molecular and physiological analysis. Annual Review of Plant Physiology 35:387-413.

38. New P.B., Kerr A., 1972. Biological control of crown gall:field measurements and glass house experiments. Journal of Applied Bacteriology 35: 279-287.

39. Nissen, P. (1971) Symposium on Informative Molecles in Biological Systems, ed. L. Ledoux, (Amsterdam,Netherlands: North Holland Publishing Co.) p. 201.

40. Ohyama, K., Pelcher, E., Schafer, A. and Fowke, L. C. (1979) Plant Physiol., 63, 382

LAP LAMBERT ACADEMIC PUBLISHING AG & CO. KG, DUDWELLER LANDSTR, GERMANY

41. Otten, L., H. DeGreve, J. Leemans, R. Hain, P. Hooykaas, and J. Schell. 1984. Restoration of virulence of vir region mutants of Agrobacterium tumefaciens strain B6S3 by coinfection with normal and mutant Agrobacterium strains. Mol. Gen. Genet. 195:159–163.

42. Pazour, G. J., and A. Das. 1990. Characterization of the VirG binding site of Agrobacterium tumefaciens. Nucleic Acids Res. 18:6909–6913.

43. Phil Oger, Annik Petit & Yves Dessaux . Genetically engineered plants producing opines alter their biological environment. Nature Biotechnology 15, 369 - 372 (1997) doi:10.1038/nbt0497-369

44. Pierronnet, A. and Salesses, G. (1996) Behavior of Prunus cultivars and hybrids towards Agrobacterium tumefaciens estimated from hardwood cuttings. Agronomie 16, 247–256

45. Powell, B. S., P. M. Rogowsky, and C. I. Kado. 1989. virG of Agrobacterium tumefaciens plasmid pTiC58 encodes a DNA-binding protein. Mol. Microbiol. 3:411–419.

46. Pu X.A., Goodman R.N., 1993. Effects of fumigation and biological control of infection of indexed crown gall free grape plants. American Journal of Ecology and Viticulture 44: 244-249.

47. Reynders-Aloisi, S. et al. (1998) Tolerance to crown gall differs among genotypes of rose rootstocks. HortScience 33, 296–297

48. Robbs, S. L., M. C. Hawes, H.-J. Lin, S. G. Pueppke, and L. Y. Smith. 1991. Inheritance of resistance to crown gall in Pisum sativum. Plant Physiol.95:52–57.

49. Schell, J. in Genetic Manipulation with Plant Material Vol. 3 (ed. Ledoux, L.) (NATO Advanced Study Institute Series, 163–181, 1975).

50. Smarrelli, J., M. T. Watters, and L. H. Diba. 1986. Response of various cucurbits to infection by plasmid-harboring strains of Agrobacterium. Plant Physiol. 82:622–624.

51. Smith, V. A. and Hindley, J. (1978) Nature (London), 276, 498.

52. Stachel, S. E., and E. W. Nester. 1986. The genetic and transcriptional organization of the vir region of the A6 Ti plasmid of Agrobacterium tumefaciens. EMBO J. 5:1445-1454.

53. Stachel, S. E., E. Messens, M. Van Montagu, and P. Zambryski.1985. Identification of the signal molecules produced by wounded plant cells that activate T-DNA transfer in Agrobacterium tumefaciens. Nature (London) 318:624-629.

54. Stapleton J.J., DeVay J.E., 1986. Soil solarization: a nonchemical approach for management of plant pathogens and pests. Crop Protection 5: 190-198.

55. Sule, S. et al. (1994) Crown gall resistance of Vitis spp. and grapevine rootstocks. Phytopathology 84, 607–611

56. Thomashow, M. F., Panagopoulos, C. G., Gordon, M. P. and Nester, E. W. (1980) Nature (London),283, 794.

57. Thomashow, M.F., Karlinsey, J.E., Marks, J.R. and Hurlbert, R.E. (1987). Identification of a new virulence locus in Agrobacterium tumefaciens that affects polysaccharide composition and plant attachment. Journal of Bacteriology 169:3209-3216.

58. Tzfira, T. and Citovsky, V. (2002) Partners-in-infection: host proteins involved in the transformation of plant cells by Agrobacterium. Trends Cell Biol. 12, 121–128.

59. Utkhede R.S., Smith E.M., 1990. Effects of fumigants and Agrobacterium radiobacter in the environment. Applied and Environmental Microbiology 59: 2112-2120.

60. Van Larebeke, N. et al. Nature 252, 169–170 (1974).

61. Van Larebeke, N. et al. Nature 255, 742–743 (1975).

62. Ward, D.V. et al. (2002) Agrobacterium VirE2 gets the VIP1 treatment in plant nuclear import. Trends Plant Sci. 7, 1–3

63. Watson, B., Currier, T. C., Gordon, M. P., Chilton, M.-D. & Nester, E. W. J. Bact. 123, 255–264 (1975).

64. Willmitzer, L., de Beuckeleer, M., Lemmers, M., van Montagu, M. and Schell, J. (1980) Nature (London), 287, 359.

65. Winans, S. C., P. R. Ebert, S. E. Stachel, M. P. Gordon, and E. W. Nester. 1986. A gene essential for Agrobacterium virulence is homologous to a family of positive regulatory loci. Proc. Natl. Acad. Sci. USA 83:8278–8282)

66. Xu, X.Q. and Pan, S.Q. (2000) An Agrobacterium catalase is a virulence factor involved in tumorigenesis. Mol. Microbiol. 35, 407–414.

67. Yadav, N. S., Postle, K., Saiki, R. K., Thomashow, M. F. and Chilton, M. D. (1980) Nature (London),287, 458.

68. Zaenen, I., Van Larebeke, N., Teuchy, H., Van Montagu, M. & Schell, J. J. molec. Biol. 86, 109–127 (1974

69. Zhu, Y. et al. (2003) Identification of Arabidopsis rat mutants. Plant Physiol. 132, 494–505.

70. Zupan, J.R. and Zambryski, P.C. (1995). Transfer of TDNA from Agrobacterium to the plant cell. Plant Physiology 107:1041.1047.

CHAPTER-7

METAL RECOVERY FROM LEAN GRADE ORES THROUGH BACTERIA

Asha Lata Singh[1] and Prakash K Singh[2]

1 Environmental Science, Department of Botany, BHU, Varanasi-221005, e-mail:

2 Coal & Organic Petrology Lab, Department of Geology, BHU, Varanasi-221005, e-mail:

ABSTRACT

The bioleaching technique has emerged as a productive and exciting field of research to extract metals from the low grade ores. This technology has gained due importance in recent years owing to its cost effective, easy scale up, up to 99% metal extraction and environmental friendly nature. For better yield of metal extraction from the low grade ores with the help of bacteria, several techniques like -In situ Bioleaching, Heap Bioleaching, Dump Bioleaching and Stirred Tank Bioleaching are being developed. Factors like pH, temperature, pulp density and nutrient media play significant role to boost up metal extraction .Various techniques involving bacterial leaching have been discussed to demonstrate extraction of copper, zinc, gold and uranium metals from the low grade ores.

Keywords: Bioleaching, ores, copper, zinc, gold, uranium

INDEX

Introduction

High grade ores containing variable concentrations of precious and non precious metals are depleting day by day at faster rate due to their increased demand in the mineral based industries. This has led to a situation where, in many cases, we are left only with low grade ores. The traditional methods of metal extraction from ores involve steps like roasting and smelting which are not only expensive but also require high concentration of metals in the ores. Moreover, they are not environmental friendly. Therefore, there is need of eco- friendly and cost effective technology for extracting metals from low grade ores. Under such a situation bioleaching (bacterial leading) is the right tool which can be appropriately used to extract metals from lean grade ores (Bosecker, 1997). This tool is becoming increasingly important alternative to the conventional process of metal extraction (DaSilva, 1981 & 1982; Gentina and Acevedo, 1985; Warhurst, 1985). The technique is quite useful and in some cases the extraction yield is up to 99%.

The bioleaching process dissolves valuable metals from lean grade ores using selective bacteria (mesophiles, moderate thermophiles and extreme thermophiles) (Romano et al., 2001). Some of the important bacteria useful in bioleaching include - Acidithiobacillus thiooxidans, Acidithiobacillus ferrooxidans, Acidithiobacillus caldus, Acidiphilium acidophilum, Leptospirillum, Archaebacteria, Acidimicrobium, Ferromicrobium, Sulfolobus spp, Acidianus, Metallosphaera, Sulfurisphaera, and Thermoplasmales. Among them, Thermoplasmales, Acidithiobacillus thiooxidans and Acidithiobacillus ferrooxidans are extremely acidophilic sulphur and iron (II) –oxidizing bacteria (Kelly and Wood, 2000). Acidithiobacillus caldus is a thermophilic leaching bacterium and is Gram-negative γ-probacteria. Acidiphilium acidophilum is a species of the genus Acidiphilum (Hiraishi et al., 1998). Members of Leptospirillum bacteria belong to a new bacterial division (Hippe, 2000; Coram and Rawlings, 2002). Metal leaching Gram positive bacteria have moderate thermophilic nature and belong to the genera Ferromicrobium, Acidimicrobium, and Sulfbacillus spp (Clark and Norris, 1996). Archaebacteria is also a leaching bacterium and belongs to the Sulfolobale; it is extremely thermophilic, sulphur and iron (II) ion-oxidizing genera like Sulfolobus, Acidianus, Metallosphaera and Sulfurisphaera (Fuchs et al., 1995, 1996; Kurosawa et al., 1998; Norris et al., 2000). Among leaching bacteria the metabolic diversity is seen according to their carbon assimilation pathways. The growth behaviour of Acidithiobacillus spp. and Leptospirillum spp. is chemolithotrophic. However, Acidiphilium acidophilum and Acidimicrobium ferrooxidans can grow autotrophically with sulphur and iron (II) compounds, and heterotrophically with glucose or yeast extract and mixotrophically with all of these substrates (Clark and Norris, 1996; Hiraishi et al., 1998). Some of the bacteria posses pigments and thus they do photosynthetic activity (Hiraishi et al., 1998, 2000; Hiraishi and Shimada, 2001).

A variety of ecological factors affect the efficiency of bio-leaching. They are temperature, pH, iron supply, oxygen and availability of other nutrients like ammonium nitrogen, phosphorus, sulfate, and magnesium etc. A generalized flowchart of bioleaching of lean grade ores (after Torma, 1977) is given in Fig 1. The present paper is focussed mainly on the role of bacteria in extraction of metals from lean grade ores.

Mechanism of metal extraction

   1.   Microbial Assimilation of heavy metals

Effluents containing elevated concentrations of heavy metals like lead, cadmium, chromium, iron, nickel, zinc in dissolved and particulate forms are being discharged from mining and industries. The physico-chemical methods involving chemical precipitation, chemical oxidation and reduction, complexation, ion exchange may be efficiently applied for removal of heavy metals. However, most of the techniques are costly and create secondary sludge problem. Under such a situation, microbes work more efficiently for uptake, adsorption and for accumulation of heavy metals from soluble as well as particulate forms of metals. This is more applicable particularly when external concentrations of metals are dilute (Shankar et al., 2007; Hussein et al., 2003). Adsorption involves metal binding on surface and is energy independent process related with both living as well as nonliving microbes. Dead and living microbial cell surface contains large number of negatively charged groups (Mullen et al., 1989).During treatment of effluents (containing heavy metals) with microbes resulting binding of positively cations with negatively charge of microbial surface through electrical attraction and cations can be removed (Fig.2).

Microbial uptake of metal is an energy dependent process which is possible using living microbes only (Charley and Bull, 1979; Norris and Kelly, 1979; Brierley, 1982). The biosorptive technique is comparatively better than the bio accumulative processes, mainly because living systems (active uptake) require supplement of nutrients which increases BOD and COD of effluents (Hussein et al., 2004).

Increase of pH increases negative charge on the cell surface favouring electrochemical attraction and metal adsorption (Gourdon et al., 1990).Bacillus, Pseudomonas sp. and micrococcus sp. have the ability to adsorb maximum Cu, Cd and Pb at pH 7, 6. Wang and Chen (2006) have also given similar results. Kim (2005) observed the growth of Bacillus sp. in elevated concentrations of heavy metal media and is capable of heavy metal adsorption. Bacillus sp.is seen to absorb the Cu at 400 mg/l and remove 65% of Cu during the active growth cycle. Pseudomonas sp. is an effective biosorbent because of its high adsorption capacity compared with Bacillus sp. and Micrococcus sp. Pseudomonas sp. is seen to adsorb 86.66% of Cd. Whereas Hussein et al 2004 have reported 88% of Cd adsorption by Pseudomonas sp. Thus Pseudomonas sp is efficient for the removal of Cu, Cd and Pb from aqueous solution. Micrococcus sp.can adsorbs 79.22% of Pb (Zaied et al., 2008). Heavy metal removal by biomass has several advantages. The processes of removal are very fast. The micro organisms have extracellular and cell wall biopolymers like polysaccharides, chitin, teichoic acid, phospholipids and proteins which provide enhanced metal immobilization and /or complexation. Heavy metals removal from biomass is reversible process and metals can be separated from biomass using various technique. Thus at present it is considered as a potential substitute for the existing technologies for effluent treatment. Huang et al., 2001 have demonstrated the removal of heavy metals from metal plating discharge using bacteria. E .coli was found to remove up to 68.5% of Cd and 58. 1% of Cr out of 48ppm of Cd and 36ppm of Cr. Bacillus sp. and S. capitis has ability to reduce hexavalent chromium to non-toxic trivalent chromium due to induced protein having molecular weight around 25 KDa. It can be removed easily from dilute solutions after reduction (Zahoor et al., 2009).

LAP LAMBERT ACADEMIC PUBLISHING AG & CO. KG, DUDWELLER LANDSTR, GERMANY

2. Bacterial leaching

2.1 In Situ bioleaching

The ores which are suitable for in-situ bioleaching generally occur below the aquifer. A number of microorganisms are known which are used for the recovery of metals from ores (table 1). The bioleaching process involves treatment of ores without mining them out to extract metals. In this process the ore is fractured by blasting or through natural process which results into development of voids and porosity in situ in ore body. This allows the free solution to flow in ore. The solution gets collected at the bottom of the mine for metal recovery. Such process can only be applied where favourable geological conditions are present with a permeable ore body and an impermeable bed rock (Fig3). This process is, however, very slow and the metal recovery is also low. This technique has the following advantages and disadvantages (**Rawlings,** 2002)

Advantages:

- Possible to mine inaccessible ore bodies
- Shorter time for mine development
- No costs for excavation
- Lower cost for mining and infrastructure
- Reduces visual environmental impact of mining operation
- Reduced radiation and other hazards

Disadvantages:

- Problems related to permeability of ores
- Impermeable ores must be cracked by explosions
- Precipitation of secondary minerals may lead permeability problems
- The leaching liquid may stream down below the ore body
- Contamination of ground water due to poor solution control

The technique has been used for copper, uranium and soluble minerals. It was discovered in the sixties that uranium could be recovered by bioleaching, subsequently industrial-scale uranium bioleaching was tried by spraying stope walls with acid mine drainage and by the in situ irrigation of fractured underground ore deposits. Such processes have also been applied to certain copper and other ore deposits. In situ bioleaching is carried out generally on the haloes of the low-grade ore that are left behind after the high-grade ores have been taken out (Rawlings, 2004).

Aspergillus foetidus has been used for nickel bioleaching from nickel bearing laterite ore bodies in situ. High salinity of water and soil in locality was seen to cause a major abiotic stress for the Ni bioleaching process by fungus Aspergillus foetidus. Hence salinity tolerant strain of fungus Aspergillus foetidus was developed (Valix et al., 2009). A mixed culture of Thiobacillus ferrooxidans, T.thiooxidans and Leptospirillum ferrooxidans has been tried in situ for leaching of Cu and Zn in the Ilba mine in Romania. After 24 months the metal output amounted to 10%Cu and 78% Zn and Cu bioleaching was observed more than Zn. The main controlling factors in the bioleaching process were high humidity and high temperature ($>20\,^0C$) (Hole , 1993).

2.2 Biooxidation heap

In this process, ore is required to be crushed and treated with sulphuric acid. It is then agglomerated in rolling drums to bind fine particles to coarse particles (Schnell, 1997) and stacked 2-10 m high on irrigation pads which are lined with high density polythene to prevent solution loss. Aeration pipes permit aeration and expedite the bioleaching process. Inorganic nutrients like $(NH_4)_2SO_4$ and $K_2HPO_4$are added on the heap surface. Leaching process of heap requires nearly one month. This technique is more useful for the treatment of lean-grade ores. The construction and operation of heap reactor are comparatively cheaper. However, the maintenance of aeration, pH gradient and nutrient is a difficult task in this technology (Fig.4). A heap bioleaching plant made in 1994 at Quabrada Blanca in Northern Chile produces 75000 tonnes of Cu per annum from Chalcocite ore containing 1.3% Cu (Schnell, 1997).

Heap reactor may also be used for gold bearing ores. For this purpose heap of agglomerated ore is initially treated with an acid ferric iron solution containing bacteria and subsequently heap reactor is operated (Brierley, 1997). The heap is washed to remove acid and cyanide consuming compounds after ore decomposition. The heap is further treated with lime, restacked on lined pads and gold is chemically extracted using a dilute solution of cyanide. The plant is installed in Carlin, Navada for gold extraction (Brierley, 1997).This technique is useful in lean grade ores (1 gm Au per tonne). Which are generally considered as waste.

In most of the heap leaching cases biooxidation occurs in the 20-35 ^{0}C range and sometimes the temperature depends on the type of microbes (Rawling,2002). Acidothiobacillus ferrooxidans, At. Thiooxidans and At. Caldus are sulphur oxidixing bacteria isolated from sulphur spring and acid mine drainage (Kelly and Wood ,2000; Lane et al., 1992; Rawling , 2001).

At. Ferrooxidans is obligate autotrophs. They are also able to grow on formic acid (Pronk et al., 1991 a). At. Ferrooxidans is able to use either ferrous iron or a variety of reduced inorganic sulphur compounds as an electron donar. It is aerobic however its also able to grow using ferric iron as an electron acceptor and reduced inorganic sulphur compound as an electron donar (Pronk, 1991 b).

At. Ferrooxidans is obligate autotrophs (Pronk et al., 1991 a) and mainly used as a dominant bacterium in heapleaching environment. This is useful in uranium and copper oxide/sulphide leaching, especially if the ferrous iron concentration in solution is high (5g/lier) (Pizarro et al., 1996). The growth of At ferrooxidans is rapid compared with other biomining bacteria in temperature range of 20-35 0C and pH 1.8-2.0. Leptospirillum ferrooxidans is more acid tolerant (pH 1.5-1.8), gram negative, and chemolithoautotrophic bacteria. Which is capable of oxidizing and using only ferrous iron .The ferrous iron oxidation is not inhibited by ferric iron. Leptospirilli form is strong iron oxidizer. Which is used in combination with sulphur oxidizing bacterium At. Caldus or At. Thiooxidans in the process of biooxidation.

LAP LAMBERT ACADEMIC PUBLISHING AG & CO. KG, DUDWELLER LANDSTR, GERMANY

2.3 Stirred tank bioleaching

This technology is useful only if the concentration of metals is sufficient to justify the cost of installing and operating the equipment. The bacterial oxidation of crushed and ground mineral slurry is carried out in aerated agitated vessels. The bioreactors are operated in continous flow mode The feed is added to the first tank and allowed to from tank to tank until biooxidation of the concentrate is complete (Lindstrom et al., 1992; Van Aswegen et al., 199. In the first stage the bioreactor is arranged in parallel. This provides sufficient retention time for the microbial cell numbers for reaching high steady state levels without being washed out (**Dew, 1995**). Mineral concentrate suspended in water is used as feed and a small quantity of fertilizer-grade $(NH_4)_2SO_4$ and KH_2PO_4 is added. Mineral decomposition occurs under pH and temperature controlled conditions. Stirred tank reactors work in the temperature range between 40 and 50°C. Oxidation rate are very high in stirred tank bioleaching as compared to in situ or heap systems. Mineral decomposition, in stirred tank reactors, takes only a few days while in heap process it takes weeks or even months for mineral decomposition. This technique gives optimum result at a pulp density of 20%. If the pulp density exceeds this limit, physical and microbial problems occur (Fig.5). Since mineral biooxidation is an exothermic process, the bioreactors are required to be cooled to remove excess heat. A large volume of air is blown through each bioreactor, and agitator is used to ensure that the solids remain in even suspension and are carried over into the next tank.

Stirred tank bioreactors are actually pre-treatment processes employed for the recovery of gold from recalcitrant arsenopyrite concentrates. In which the gold is finely divided in a mixture of pyrite/arsenopyrite and cannot easily be solubilized by the cyanidation process. Ores are treated to decompose the arsenopyrite to allow the cyanide to make contact with the gold. Since gold occurs in small fraction, the ore is crushed and a gold containing concentrate is prepared by flotation. Which makes up 18-20% of the feed volume and is fed at a rate that permits a total residence time of about four days in the series of stirred tanks (Dew, 1997).The microbes release ferric iron and sulphate, which eventually acidifies the environment, and the solution is maintained at pH 1.5-1.6.

Eric Livesey-Goldblatt recognised the potential of microbes for the treatment of these lean grade high value minerals during 1980s (Livesey et al., 1983)). While many biooxidation plants were built by Gencor which operated at 40 ^{0}C (Dew, 1995,1997), the plant at the Youanmi mine in Western Australia designed by Bac Tech, operated at 50 ^{0}C (Miller PC.,1997).After required amount of mineral biooxidation is completed the recovered mineral is washed and separated from the liquid fraction in thickeners. It is further neutralized, and the gold is extracted with cyanide. In 1986, the first commercial operation was commissioned at the Fairview mine in Barberton, South Africa to use stirred tank biooxidation of gold. Since then several plants have come up in other countries, including Australia, Brazil, Ghana and Peru (Brierley CL, 1997). The plant at Sansu, Ghana expanded in 1995, is operating one of the largest fermentation process in the world and processes 1000 tonnes of gold concentrate per day

Stirred tank processes have also been developed for the ores, which include commercial scale cobalt bioleaching plant initiated by BRGM of France at Kasese, Uganda (Briggs et al., 1997) and the pilot scale BioNIC process for leaching of nickel by Billiton Process Research in South Africa (Dew and Miller 1997).New plants to recover copper from chalcopyrite ore and operating at temperatures over of 75 ^{0}C are also being tested. The bacteria used in stirred tank processes include At.ferrooxidans, At.thiooxidans,At. caldus and Leptospirillum etc.

2.4 Vat bioleaching

Rectangular containers mostly drums, barrels, tanks or vats are used as units in vat bioleaching technique. These containers are made up of concrete or wood and are internally lined with resistant material to the leaching media. Generally coarse ores are used in this technique. Vats are operated sequentially to get more contact period between the ores and the reagents. Here, dissolution of crushed ores is carried out in a tank with uninterrupted stirring. The leachate collected from one container is added to another vat with fresh ores. Thus, the yield increases due to 'Counter current extraction' which means leachate and ore move counter-currently. Thus the metal recovery can be enhanced under controlled condition in this technique. Though the process is expensive, it is preferred for ore concentrates and precious metals.

2.5 Dump leaching

This process uses uncrushed ores which are piled up and contain around 0.1-0.5% Cu .The conventional methods are not suitable because of low metal content and no chance of profitability. Some of the dumps are huge and may accommodate around 10 million tons of waste rock (very lean grade ores). Lixiviant are applied to the top of the dump surface, in case of dump leaching, and recovery of the metal laden solution gets at bottom which comes through gravity flow. The dilute sulphuric acid added at the top gradually percolates down through the dump. This helps in decreasing the pH value and enhances the growth of acidophilic microorganism. Acid is collected at the bottom of the dump. This process is very slow and takes much time in recovery of metals from the ores.

Bio-leaching of metals

1. Copper Bio-leaching

In general, most of the ore bodies of copper are mined from minerals resulting from the weathering of the primary copper ore mineral chalcopyrite [$CuFeS_2$]. The minerals occurring in the enriched zone are chalcocite [Cu_2S], bornite [$2Cu_2S \bullet CuS \bullet FeS$] and djurleite [$Cu_{31}S_{16}$], while the minerals in the oxidized zones are malachite [$CuCO_3 \bullet Cu$ (OH) $_2$], azurite [Cu_3 $(CO_3)_2(OH)$ $_2$], chyrsocolla [$CuO \bullet SiO_2 \bullet 2H_2O$], cuprite [$Cu_2O$], tenorite [$CuO$], native copper and brochantite [$CuSO_4 \cdot 3Cu$ (OH) $_2$].

Countries like Chile, Indonesia, Mexico, Peru, and Zambia are sharing 50% of the global copper production and nearly 25% of the global copper production comes from bioleaching process. Copper is mainly recovered by heaps and dumps bioleaching process

(Table 2). The process of bioleaching converts water insoluble copper sulphide into water soluble copper sulphide form. During heap bioleaching minerals like chalcocite or covellite are firstly crushed, acidified by using sulphuric acid. Secondly it is agglomerated in rotating drums to bind fine materials to courser particles (Rawlings 2002).Thirdly, stacking in heaps and treated with iron containing solution. This solution gets into the heap and bacteria. Finally bacteria catalyze the reaction and copper released.

Thus bioleaching was considered as an appropriate technique during the period of 1950 to 1980 for extraction of copper and other metals from low grade ores. During these years the bioleaching plants were operated for the recovery of copper from the Copper Mines of Rio Tinto, Spain and in Cananea, Mexico. Nearly 8000 tonnes of copper were produced per year in the mines of Rio Tinto and 9000tonnes of copper in Cananea (Gentina and Acevedo, 1985). RioTinto took the lead in establishing the first commercial scale bioleaching operation for the recovery of copper from the Copper Mines in Spain (Brierley, 1978, 1982). Subsequently, in mid eighties, a major breakthrough came in Minera Pudahuel (Chile) when heap bioleaching was successfully applied to recover nearly 14000 tonnes of copper per year from the lean ores containing 1-2 % copper (Acevedo et al., 1993)(Table 3).

Flotation can separate minerals on the basis of differences in surface properties. In the flotation process hydrophobic mineral particles get attached to the air bubbles and move to the top of the flotation cell. Whereas, the hydrophilic surface particles get wet and sink to the bottom of the cell and form tailings. Study of the chemical and physico-chemical properties of mineral surfaces is important in the floatation system to know the effect of changes in composition of bulk phases (Luttrell and Yoon, 1984). Bacterium like A. ferrooxidans is seen to prevent the flotation of pyrite (Attia and El-Zeky, 1989; Santhiya et al., 2000).

Yuce et al., (2006) studied the copper grade of flotation concentrate treated with bacterial conditioning and realized an increase of 22 % in copper concentrate while the conventional flotation gave only 18-20% increase of Cu concentrate.

Bacterial leaching of Cu sulphide flotation concentrate is a complex process. Sadowski et al., (2003) investigated through bioleaching using flotation technique and achieved high Cu extraction in the presence of 3% pyrite. However, when pyrite limit exceeded 3 %, it inhibited the process.

Tipre and Dave (2004) used Acidithiobacillus ferrooxidans strains for leaching of Cu from low grade Cu ore using stirred tank reactor and studied the role of pH and pulp density. They observed a maximum leaching at pH 2.5 and pulp density 20% and beyond this limit the Cu leaching decreased.

Acidithiobacillus ferrooxidans is a widely used bacterium for bioleaching of sulphide minerals. Heavy metal ions get accumulated in the leach liquor. It may become toxic to the organisms after certain concentration and affect dissolution rates. Metal tolerant strains of Acidithiobacillus ferrooxidans are therefore important for efficient leaching. Such strains have been used in the bioleaching of Sungai Lembing tin mine (Malayasia) by product which gave a better result (Natarajan et al., 1994; Elzehg and Attia, 1995; Das et al., 1997).

2. Zn bio-leaching

The chief zinc ore mineral is sphalerite [ZnS]. However, some zinc production has also been made from smithsonite [$ZnCO_3$] and hemimorphite [$Zn_4Si_2O_7$ $(OH)_2$ $\cdot H_2O$]minerals in the past.

Deveci et al., (2004) carried out bioleaching process of McArthur river ore (Australia) for extraction of zinc. They used mesophilic, moderately thermophilic and extremely thermophilic strains of acidophilic bacteria. Strains of A. ferrooxidans (mesophilic), Sulphobacillus acidophilus and Sulphobacillus lostonensis (moderalrly thermophilic) and Acidianus brierleyi (extremely thermophilic) were grown and maintained on the ore. They have demonstrated that the moderately thermophilic bacteria - Sulphobacillus lostonensis, is more efficient for bioleaching of zinc than the extremely thermophilic A. brierleyi. Moreover, increase in acidity at pH less than 1.6, inhibited the bioleaching capacity of mesophiles and moderate thermophiles. However, A. brierleyi could tolerate higher acidity (pH 1-2). Iron treatment was found as an enhancer for bioleaching of zinc from their ores. Ghosh et al., (2004) also showed a positive effect of sulphuric acid and could extract 100 % zinc from the Sikkim (India) ore.

Hossain et al., (2004) carried out bioleaching of Zn Sulfide ore using Thiobacillus ferrooxidans under aerobic condition and demonstrated that maximum bioleaching of Zn occured at 74.85 % (w/w) and 71.50% (w/w) for 25 kg/m^3 and 30kg/ m^3 pulp density respectively. They performed the experiment with the ore size between -200 and + 240 mesh at pH 2.5 and temperature $35^{\circ}C$ and determined maximum leaching at 3 % (w/w) glucose and 0.3% nitrogen concentration respectively. The tolerance of bacteria is up to 40kg/ m^3 initial Zn sulphide loading. In addition to bioleaching process, the 'stirred tank reactors' and 'bioreactors' are also being employed in several countries to recover zinc. Bioreactor is proved to be more useful than heap and dump bioleaching process because it provides homogenous reacting mass and enhancement in metal extraction rates (Acevedo, 2000). Tipre and Dave, (2004) have used consortium of Acidithiobacillus ferrooxidans, Thiooxidans, Leptospirillum ferrooxidans and hetrotrophic organism for Cu, Pb and Zn bioleaching at different pulp densities. It was observed that at 20 % (w/v) pulp density the metal extraction rate is optimum. Higher pulp density inhibits growth of bacteria and rate of ferrous oxidation (Bailley and Hansford, 1993).

3. Gold bio-leaching

Though the primary mineral of gold is the native metal but some tellurides such as calaverite [$AuTe_2$], sylvanite [(Ag, Au) Te_2], and petzite [Ag_3 Au Te_2] are important ore minerals. Due to depletion of gold from surface deposits, it is increasingly being mined from deeper levels, where it occurs intermixed with and enclosed by sulphide minerals like chalcopyrite (Ca Fe S_2) and pyrite (FeS_2). It is very difficult to extract gold from these lean grade ores, by sodium cyanide leaching process (i.e. chemical leaching).

The sulfide gold ores are pre-treated by roasting or by pressure-oxidation to free the gold from the enclosing sulfides prior to cyanide leaching. The pre-treatments are very costly in the case of low grade ores. Therefore, substantial savings may be done by substituting these costly pre-treatment by bio-leaching. This is being followed in S. Africa, Ghana, Australia, USA, Brazil etc (Table 4). Bio heap and stirred tank technique is mainly used for the recovery of gold (Rawlings 2004).

T. ferrooxidans, T. thiooxidans, and Leptospirillum ferrooxidans are more efficient for the extraction of gold particles from the sulphide minerals. Thermophilic Sulfolobus strain also plays a role in dissolution of sulphide minerals. Among these bacteria, T ferrooxidans is frequently used in the leaching of gold bearing sulphides. It is gram negative, rod shaped, acidophilic, motile, non-flagellated, non- spore forming and obtains energy by the oxidation of sulphur and iron. Its optimum growth can be obtained at pH 2-3 and temperature between 30 and 35°C. Bacteria under the Aeromonas genus are also able to dissolve gold at pH 8-8.5. Bacillus subtiles and B. mescentericus solubilize gold by secreting gold solubilizing peroxides and amino acids.

At different gold processing plants, several million tonnes of lean grade ores (0.2 to 1 ppm of Au) are present. Bio heap leaching is the best technique to extract gold from such lean grade ores. In S. Africa, the tailing of slime dam has been adopted for bio-treatment. In first step, the ploughing up of compact layer of tailing is done to convert loose granular mass to allow air, water and inoculation of bacteria. The ploughing of compact layer of tailing is processed and bacteria liberate the trapped gold particles from the waste dumps.

Some bacteria are highly efficient for the enrichment of gold by two processes - absorption and adsorption. In absorption, the bacteria accumulate gold inside their cells and bind with the mercepto group of protein. However, in adsorption, the binding of gold takes place by amino acids secreted by cell wall of the organism. Thus, it may act as an indicator in the investigations of gold bearing rocks. Proteus vulgaris (aerobic), Bacillus subtilis (aerobic) and micrococcus have been used for gold enrichment by Zingrong et al., (1997) who realized that adsorption of gold is more than absorption. Protein rich organs of animals and plants also accumulate gold (Boyle, 1990). Absorption of trace gold enhances the metabolic and other biochemical functions of bacteria, because it combines with –HS and forms metallic protein (Huang Shuhui, 1990).However, adsorbed gold on bacterial surface may combine with amino acid, polysaccharides and other organic species. Some of the researchers suggest that gold combine with amino acid is mainly coordinated with atomic nitrogen in the amino group to form co- valent bounded complex and oxygen in the carboxyl group to form ionic bounded complex.

In future, it is required to develop metal tolerant strains to enhance the metal extraction from lean grade ores. Metal tolerance strains are more efficient for metal extraction from lean grade ores than the non-metal tolerance strains. The reason behind this is the metal non-tolerant strains get inhibited by the toxicity of leached metals.

4. Uranium bio-leaching

The main ore minerals are uraninite [UO_2], pitchblende (a mixture of various oxides), coffinite [U (SiO_4) 1-x (OH) 4x] and a host of secondary minerals such as carnotite [K_2 (UO_2)$_2$ (VO_4)$_2 \cdot 3H_2O$] and autunite [Ca (UO $_2$) $_2$ (PO_4) $_2 \cdot$ 10-12H_2O].

Uranium is yet another metal that can be successfully recovered using microbial leaching (Brierly, 1982). This process can be commercially utilized for lean-grade uranium deposits as well as for the recovery of uranium from low grade nuclear wastes (Fig.6).

The tetravalent uranium oxides (UO_2) that occur in lean grade ores are insoluble. It can be converted to a leachable form by oxidizing it in ferric (Fe^{3+}) ions. The chemical oxidants are highly expensive and add much cost to the leaching process. When the uranium ores are associated with pyrite, the $Fe3^+$ oxidant is produced and may be regenerated at lower cost using bacteria like T. ferrooxidans (Guay et al., 1977). Bacteria like T. thiooxidans (sulphur oxidizer) and Leptospirillum ferroxi (iron oxidizer) use CO_2 as the sole source of carbon and grow in the acidic pH range of 1-3 which is typically used in uranium leached solution. However, it was observed that T. ferrooxidans is the most important bacterium for microbiological leaching of uranium ores. Using Acidithiobacillus ferrooxidans, Sung et al (2005) could extract 80 % uranium from the lean grade black schist from Okcheon area of S. Korea within 60 h and at a pulp density of 100 g-ore/l. The leaching efficiency could be maintained even at a high pulp density of 500 g-ore/l.

The recovery of uranium using thiobacillus depends on the composition of the mineral deposit. The pyritic uranium oxide ores are ideal for such bio-leaching as compared to the ores associated with carbonates. The bacterial leaching of uranium is generally feasible in those cases where the ores are in the tetravalent state and are associated with reduced sulfur and iron minerals (Atlas and Bartha, 2005). Besides, there are other factors that affect the rate of uranium bioleaching like morphology and electronic structure of the mineral surface, pH, redox potential, particle size, surface area, temperature, partial pressure of oxygen, relative humidity and climatic condition.

Conclusions and Future prospects

Bioleaching is an innovative technique used for the recovery of a number of metals from their ores. Application of traditional methods like smelting and roasting are high energy demanding and necessitate high grade ores. However, bioleaching has potential to extract metals from lean grade ores and require less energy input. Besides, these processes are environmental friendly and give high extraction yield. Depending on the suitability, 'insitu bioleaching', 'vat bioleaching', 'dump bioleaching', 'stirred tank bioleaching' and 'heap bioleaching' may be employed. Several species of Thiobacillus genus have been successfully employed to recover copper, zinc, gold, and uranium. Spain, Mexico, Chile, Indonesia, Peru, and Zambia have used bioleaching to recover copper and nearly 25 % of global copper production comes through this. Substantial savings have been realized through bioleaching to recover gold in S. Africa, Ghana, Australia, and USA.

Pseudomonas japonica and Staphylococcus aureus can be used productively for the recovery of Zinc from dilute solution (Singh, 2010, Singh et al., 2008). Removal of Cr (VI) from waste water has been carried out using Bacillus mycoides in the presence of sulphate under aerobic condition (Singh et al 2006). Singh et al., (2000) have reported that resistance strains of Pseudomonas aeruginosa, could successfully remove 100 % copper from water sample that contained only 10 ppm Cu concentration. It will be very interesting to see the enhancement in extraction of metals from lean grade ores using resistant strains of bacteria developed through adaptation as well as through genetic modification.

Acknowledgement

The authors thank their respective Heads of the Departments (Department of Botany, BHU and Department of Geology, BHU) for extending the necessary facilities. The authors also acknowledge their little daughter- Anjali Singh, who always helped them by putting critical questions indirectly related to the subject.

Table 1. Iron and sulphur oxidizing bacteria (from Walting, 2006)

Organism	Reported growth substrates	Characteristics
Acidianus ambivalens	Soxidation and reduction	Hyperthermophiles
Acidianus brierleyi	Sulphides	pH opt.1.5-2.5
Acidianus infernus	Poor,if any,Fe oxidation	
	Mixotroph	Moderate thermophile pH opt.2
Acidimicrobium ferrooxidans	Fe oxidation and reduction	
	Sulphides (poor)	
Acidiphilum spp	Obligate heterotrophs	Mesophiles
Acidiphilum SJH	S oxidation	pH opt.2-3
	Fe (III) reduction	
Acidiphilum acidophilum	Facultative autotroph S oxidation Fe (III) Reduction	Mesophile pH opt.2-3
Acidithiobacillus albertensis	Autotrophs	Mesophiles
Acidithiobacillus ferrooxidans	Soxidation,sulphides	pH range 2-4
Acidithiobacillus thiooxidans	(Af,Fe(II) oxidation;Fe (III) reduction as a facultative anaerobe	
Acidithiobacillus caldus	Mixotroph 3S Oxidation,sulphides	Moderate thermophile pH opt.2-2.5
Acidolobus aceticus	Hetrotroph Sreduction to H$_2$S	Hyper thermophile pH opt.3.8
Alicyclobacillus spp	S Oxidation,sulphides	Mesophiles - Moderate thermophile
Alicyclobacillus disulfidooxidans	(Ad, facultative autotroph	pH 1.5-2.5
Alicyclobacillus tolerans	At,mixotrphs,Fe(III) reduction	
Leptospirilum thermoferrooxidans	Pyrite	pH 1.6-1.9
Leptospirilum ferrooxidans	Fe oxidation,pyrite	Mesophile
Sulfobacillus acidophilus	Fe(II) oxidation, Fe(III) Reduction, Sulphides	Moderate thermophiles
Sulfobacillus thermosulfidooxidans	S Oxidation	pH 1-2.5

LAP LAMBERT ACADEMIC PUBLISHING AG & CO. KG, DUDWELLER LANDSTR, GERMANY

Sulfobacillus metallicus	Strict Chemolithoautotrophs	Hyperthermophiles
Thiobacillus prosperus	S and Fe Oxidation Sulphides	Mesophile,halphile pH opt 2

Table 2. Heap bioleaching of copper ores (taken from Watling, 2006)

Region/mine	Operation reserves (t)	Ore processes (t/day)	Cu production (t/year)
Lo Aguirre, Chile 1980-1996	Heap bioleach $12x10^6$at1.5%Cu	Oxides/chalcocite $16x10^3$	$14-15x10^3$
Cerro Colorado, Chile 1993	Heap bioleach $80x10^6$ at 1.4%Cu	Chalcocite,covellite $16x10^3$	$100x10^3$
Ivan Zar, Chile 1994	Heap bioleach $5x10^6$ at 2.5%Cu	Oxides/sulphides $1.5x10^3$	$12x10^3$
Quebrada Blanca, Chile 1994	Heap/dump bioleach $85x10^6$ at 1.4% Cu $45x10^6$ at 0.5%Cu	Chalcocite $17.3x10^3$	$75x10^3$
Punta del Cobre, Chile 1994	Heap (bio)leach $10x10^6$ at 1.7% Cu	Oxides/sulphides	$7-8x10^3$
Andacollo Chile 1996	Heap/dump bioleach $32x10^6$ at 0.58% Cu	Chalcocite $15x10^3$	$21x10^3$
Dos Amigos, Chile 1996 -	Heap bioleach 2.5%	Chalcocite $3x10^3$	-
Zaldivar, Chile 1998-	Heap/dump bioleach $120x10^6$ at 1.4% Cu $115x10^6$ at 0.4%Cu	Chalcocite $20x10^3$	$150x10^3$
Lomas Bayas, Chile 1998	Heap/dump bioleach $41x10^6$ at 0.4% Cu	Oxides/sulphides $36x10^3$	$60x10^3$
Cerro Verde, Peru 1997	Heap bioleach at 0.7%Cu	Oxides/sulphides $32x10^3$	$54.2x10^3$
Escondida, Chile	Heap bioleach $1.5x10^9$ at 0.3-0.7%Cu	Oxides/sulphides	$200x\ 10^3$
Lince II Chile,1991	Heap leach 1.8%Cu	Oxides/sulphides	$27x\ 10^3$
Toquepala, Peru	Heap leach	Oxides/sulphides	$40x\ 10^3$
Morenci,Arizona 2001	Mine for leach $3450x\ 10^6$ 0.28%Cu	Chalcocite pyrite $75x10^3$	$380x10^3$
Equatorial Tonopah, Nevada,2000-2001	Heap bioleach at 0.31%Cu	$25x10^3$	$25x10^3$

LAP LAMBERT ACADEMIC PUBLISHING AG & CO. KG, DUDWELLER LANDSTR, GERMANY

16. Gunpowder Mammoth Mine,Australia,1991	In situ bio leach 1.2x10⁶ at 1.8%Cu	Chalcocite and bromite-	33x10³
Girilambone, Australia 1993-2003	Heap bioleach at 2.4 %Cu	Chalcocite and Chalcopyrite 2x 10³	14x10³
Nifty Copper, Australia,1998	Heap bioleach at 1.2 %Cu	Oxides/ Chalcocite 5x10³	16x10³
Whim Creek and Mons Cupri, Australia 2006	Heap bioleach 900x10³ at 1.1% Cu 6x 10⁶ at 0.8 %Cu	Oxides/sulphides	17x10³
Mt Leyshon, Australia 1992-1997	Heap bioleach-0.15%	Chalcocite 1.3x10³	750
S and K Copper Monywa,Myanmar, 1999	Heap bioleach 126x10⁶ at 0.5% Cu	Chalcocite 18x10³	40x 10³
Phoenix deposit, Cyprus,1996	Heap bioleach 9.1x 10⁶ at 0.78% Cu 5.9x10⁶ at 0.31% Cu	Oxides/sulphides -	8x 10³
Jinchuan Copper, China 2006	240x10⁶ at 0.63% Cu	Chalcocite,Covellite, enargite	10x 10³

Table 3 World copper production (thousands of tonnes)

Country	1998	1999	2000
Australia	607.0	711.0	829.0
Canada	705.8	620.1	634.2
Chile	3,6869	4,391.6	4,602.0
China	486.0	520.0	588.5
Indonesia	809.1	790.3	1,005.5
Mexico	384.3	381.2	344.6
Peru	483.3	563.3	553.9
Poland	436.2	463.6	463.2
Russia	518.0	510.0	510.0
United States	1,860.0	1,601.0	1,480.0
Zambia	378.8	271.0	320.1
Others	1,932.7	1,915.9	1,912.7
World	12,288.1	12,712.0	13,243.7

The data is from USGS Minerals Yearbook and available at http://minerals.usgs.gov/minerals/pubs/commodity/myb

(Fernando Acevedo, 2002)

LAP LAMBERT ACADEMIC PUBLISHING AG & CO. KG, DUDWELLER LANDSTR, GERMANY

Table 4. World gold production (thousands of tonnes)

Country	1998	1999	2000
Australia	310,1	310.1	296.4
Brazil	49.6	52.6	52.0
Canada	165.6	157.6	153.8
Chile	45.0	45.7	54.1
China	178.0	173.0	180.0
Ghana	72.5	79.9	72.1
Indonesia	124.0	127.1	124.6
Papua New Guinea	64.1	61.3	74.0
Peru	94.2	128.5	132.6
Russia	114.9	125.9	140.0
South Africa	464.7	451.3	430.8
United states	366.6	341.0	353.0
Uzbekistan	80.0	85.0	85.0
World	2,510.0	2,550.0	2,550.0

The data is from USGS Minerals Yearbook and available at

http://minerals.usgs.gov/minerals/pubs/commodity/myb (Fernando Acevedo,2002)

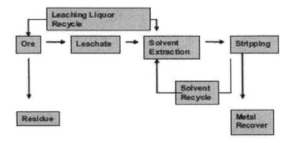

Fig.8.1 Flow chart of bioleaching by T.ferroxidans for metal recovery from lean grade ores

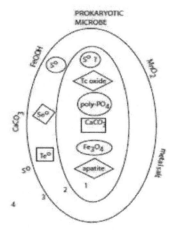

Fig. 8.2 Sites of mineral deposition in prokaryotic microbes.1.Cytoplasm (intracellular);2. Periplasm; 3. Cell Surface; 4. Bulk phase.Intracellular So? has uncertainty of its location in Achromatium oxaliferum. The So at the cells surface refers to sulfur deposited extracellularly by Chlorobiaceae, which according to Van Gemerden (1986) remains attached to the cell surface.

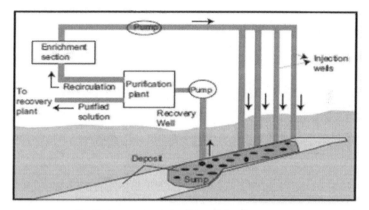

Fig 8.3 A Generalized sketch of insitu bioleaching

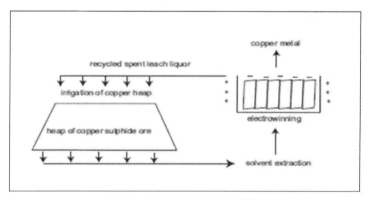

Fig. 8.4 Heap leaching of copper containing ore

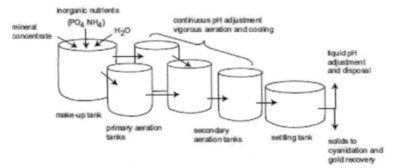

Fig 8.5 Flow diagram of a continuous flow biooxidation facility for the pretreatment of gold bearing arseno pyrite concentrate through stirred tank bioleaching

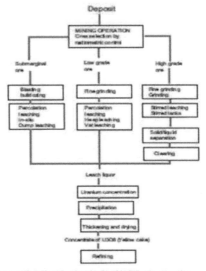

Fig 8.6 Flow Sheet for the treatment of Uranium ores

References

1. Acevedo F (2002) Present and future of bioleaching in developing countries. E J B Electronic Journal of Biotechnology.15:1-6.

2. Acevedo F (2000) The use of reactors in biomining processes. Electron. J. Biotechnol. 3:1-11.

3. Acevedo F, Gentina JC, Bustos, S. (1993) Bioleaching of minerals- a valid alternative for developing countries. Journal of Biotechnology. 31: 115-123.

4. Atlas R.M., Bartha R (2005) Microbial Ecology: Fundamentals and Application. 4th Edition. Pearson Education. 694p.

5. Attia YA, El- Zeky, M (1989) Bioleaching of gold pyrite tailings with adapted bacteria. Hydrometallurgy. 22:291-300.

6. Bailley AD, Hansford GS (1993).Factors affecting biooxidation of sulphide minerals at high concentration of solids-a review. Biotechnol.Bioeng. 42:1164-1174.

7. Bosecker K. (1997) Bioleaching: metal solubilization by microorganisms. FEMS Microbiology Review. 20:591-604.

8. Boyle M. (1990) Biodegradation of land applied sludge. Journal Environ. Qual. 19:640-644.

9. Brierley CL (1978) Bacterial leaching .CRC critical Reviews in Microbiology. 6:207-262.

10. Brierley CL (1982) Microbiological mining. Scientific American, August. 247: 44-54.

11. Brierley JA (1997) Heap leaching of gold bearing deposits: theory and operational description.72:103-115.

12. Brierley CL (1997) Mining biotechnology : research into commercial development and beyond.72:3-17.

13. Briggs AP, Millard M (1997) Cobalt recovery using bacterial leaching at the Kasese project,Uganda.In International Biohydrometallurgy Symposium,IBS97,pp.M2.4.1-2.4.12.Glenside,South Aust.:Aust.Mineral Found.

14. Charley RC, Bull AT (1979) Bioaccumulation of silver by a multispecies community of bacteria. Arch Microbiol. 123: 239-244.

15. Clark DA, Norris PR (1996) Acidimicrobium ferrooxidans gen.nov., sp.nov.:mixed culture ferrous iron oxidation with Sulfobacillus species. Microbiology.142:785-790.

16. Coram NJ, Rawlings DE (2002) Molecular relationship between two groups of the genus Leptospirillum and the finding that Leptospirillum ferriphilum sp.nov.dominates South African commercial biooxidation tanks that operate at 40 degrees C.Appl Environ Microbiol. 68:838-845.

17. Das A, Modak JM, Natarajan KA.(1997) Technical note studies on multi metal ion tolerance of Thiobacillus ferrooxidans. Mineral Engineering. 7:743-749.

18. Dasilva EJ.(1981) The renaissance of biotechnology: man, microbe, biomass and industry.Acta Biotechnologica. 1:207-246.

19. Dasilva EJ. (1982) The world of microbes. Impact of science and society. 32:125-132.

20. Deveci H, Akcil A, and Alp I. (2004) Bioleaching of complex zinc sulphides using mesophilic and thermophilic bacteria: comparative importance of pH and iron.Hydrometallurgy;73:293-303.

Dew DW (1995) Comparison of performance for continuous bio-oxidation of refractory gold ore flotation concentrates. In Biohydrometallurgical Processing, ed. T Vargas, CA Jerez, JV Wiertz, H Toledo, 1:239-51. Santiago:Univ.Chile Press.

21. Dew DW, Lawson EN, Broadhurst JL (1997) The BIOX process for biooxidation of gold –bearing ores or concentrates.72:45-80

22. Dew DW, Miller DM (1997) The BioNIC process, bileaching of mineral sulphide concentrates for the recovery of nickel. In International Biohydrometallurgy Symposium,IBS97,pp.M7.1.1-7.1.9.Glenside,South Aust.:Aust.Mineral Found.

23. Ehrlich HL (1999) Microbes as Geologic Agents: Their Role in Mineral Formation. Geomicrobiology Journal.16:135-153.

24. Elzehg M, Attia YA. (1995) Effect of bacterial adaptation on kinetics and mechanism of bioleaching ferrous sulphides. The Chemical Engineering Journal. 56: B115-B124.

25. Fuchs T, Huber H, Teiner K, Burggraf S, Stetter K. (1995) Methallosphaera prunae sp.nov., a novel metal –mobilizing , thermoacidophilic archaem, isolated from a uranium mine in Germany.Syst Appl Microbial. 18:560-566.

26. Fuchs T, Huber H, Burggraf S, Stetter K. (1996) Rdna –based phyloeny of the archaeal order Sulfolobales and reclassification of Dasulfurolobus ambivalens as Acidianus ambivalens comb nov., Syst Appl Microbiol. 19: 56-60.

27. Gentina JC, Acevedo F. (1985) Microbial ore leaching in developing countries. Trends in Biotechnology. 4; 86-89.

28. Ghosh M K, Sukla L B, Misra V N. (2004) Cobalt and zinc extraction from Sikkim complex sulphide concentrate. Trans Indian inst. Met. 57: 617-621.

29. Gourdon R, Bhande S, Rus E, Sofer SS (1990) Comparison of cadmium biosorption by gram positive and gram negative bacteria from activated sludge. Biotechnol Lett. 12:839-842.

30. Guay R, Silver M, Torma AE. (1977) Ferrous iron oxidation and uranium extraction by Thiobacillus ferrooxidans. Biotechnology and Bioenineering. 19:727-740.

31. Hippe H. (2000) Leptospirillum gen .nov. (Markosyan 1972), nom.rev.,including Leptospirillum ferrooxidans sp.nov (Markosyan 1972),nom. rev. and Leptospirillum thermoferrooxidans sp.nov. (Golovacheva et al .1992).Int J Syst Evol Microbiol. 50:501-503.

32. Hiraishi A, Nagashima KV, Matsuura K, Shimada K, Takaichi S, Wakao N, Katayama Y. (1998) Phylogeny and photosynthetic features of Thiobacillus acidophilus and related acidophilic bacteria: its transfer to the genus Acidiphium as Acidiphilium acidophilum comb.nov.Int.J Syst Bacteriol. 48:1389-1398.

33. Hiraishi A, Shimada K (2001) Aerobic anoxygenic photosynthetic bacteria with zinc –bacterio chlorophyll. J. Gen. Appl. Microbiol.47:161-180.

34. Hiraishi A, Matsuzawa Y, kanbe T, Wakao N. (2000) Acidisphaera rubrifaciens gen.nov.,sp.nov., an aerobic bacteriophyll-containing bacterium isolated from acidic environments. Int. J. Syst Evol Microbiol . 50:1539-1546.

35. Hole J (1993) Bioleaching Processes.Biohydrometallurgical Technologies.1:15-27.

36. Hossain SM, Das M, Begum KMMS, Anantharaman N. (2004) Bioleaching of Zn sulphide (Zns) ore using Thiobacillus ferrooxidans. IE (I) Journal –CH. 85:7-11.

37. Huang S. (1990) Microorganic leaching and recovery of gold: Bulletin of Microbiology. 5 :300-303 (in Chinese).

38. Huang MS, Pan J, Zheng LP (2001) Removal of heavy metals from aqueous solutions using bacteria. Journal of Shanghai University. 5(3)253-259.

39. Hussein H, Farag S, Moawad H (2003) Isolation and characterization of Pseudomonas resistant to heavy metals contaminants. Arab J. Biotechnol.7:13-22.

40. Hussein H, Ibrahim SF, Kandeel K, Moawa H (2004) Biosorption of heavy metals from waste water using Pseudomonas sp. Electronic J.Biotechnol.7(1).

41. Jingrong Z, Jianjun L, Jianping Z, Fan Y. (1997) Simulating experiments on enrichment of gold by bacteria and their geochemical significance. Chinese Journal of Geochemistry. 16: 369-373.

42. Kelly DP,Wood AP(2000) Re-classification of some species of Thiobacillus to the newly designated genera Acidothiobacillus gen.nov., Halothiobacillus gen. nov. and Thermithiobacillus gen.nov.Int.J.Syst.Evol.Microbiol.50:511-516.

43. Kelly DP, Wood AP. (2000) Reclassification of some species of Thiobacillus to the newly designated genera Acidithiobacillus gen.nov. Halothiobacillus gen.nov. and Thermithiobacillus gen.nov.Int J.Syst Evol Microbiol. 50: 511-516.

44. Kim BM (2005) In: AICHE Symposium series water. American Institute of chemical engineers. New York, 77:39-48.

45. Kurosawa N, Itoh YH, Iwai T, Sugai A., Uda I., Kimura N, Horiuchi T, Itoh T. (1998) Sulfurisphaera ohwakuensis gen. nov., sp.nov., a novel extremely thermophilic acidophile of the order Sulfolobales. Int .J.Syst Bacteriol. 48: 451-456.

46. Lane DJ, Harrison AP, Stahl D, Pace B, Giovannoni SJ (1992) Evolutionary relationship among sulphur and iron oxidizing bacteria J.Bacteriol.174:269-278.

47. Lindstrom EB,Gunneriusson E, Tuovinen OH (1992) Bacterial oxidation of refractory ores for gold recovery. Crit. Rev. Biotechnol. 12:133-155.

48. Livesey Goldblatt E, Norman P, Livesey- Goldblatt DR(1983) Gold recovery from arsenopyrite/pyrite ore by bacterial leaching and cyanidation. In Recent Progress in Biohydrometallurgy,ed.G   Rossi,AE   Torma,pp.627-41.Iglesias,Italy:Assoc.Mineraria Sarda.

49. Luttrell GH, Yoon RH. (1984) Surface studies of the collectorless flotation of chalcopyrite.Colloids and Surfaces. 12: 239-254.

50. Miller PC (1997) The design and operating practice of bacterial oxidation plant using moderate thermophiles (the Bac Tech process) 72:81-102.

51. Mullen MD, Wolf DC, Ferris FG (1989) Bacterial sorption of heavy metals. Applied and Environmental Microbiology.55 (2): 3143-3149.

52. Natarajan KA, Sudeesha K., and Ramananda Rao, G. (1994) Stability of copper tolerance in Thiobacillus ferrrooxidans. Antonie Van Leeuwenhock. 66: 303-306.

53. Norris PR, Kelly DP. (1979) Accumulation of metals by bacteria and yeasts. Dev.Ind.Microbiol. 20: 299-308.

54. Norris PR, Burton NP, Foulis NAM(2000)Acidophiles in bioreactor mineral processing.Extremophiles. 4:71-76.

55. Pizarro J, Jedlicki E, Orellana O, Romero J, Espejo RT(1996) Bacterial populations in samples of bioleached copper ores as revealed by analysis of DNA obtained before and after cultivation. Appl.Environ.Microbiol.62:1323-1328.

56. Pronk JT, Meijer WM, Haseu W,Van Dijken JP, Bos P, Kuenen JG (1991)Growth of Thiobacillus ferrooxidans on formic acid. Appl.Environ.Microbiol.57:2057-2062.(a)

57. Pronk JT, Liem K, Bos P, Kuenen JG (1991) Energy transduction by anaerobic ferric iron respiration in Thiobacillus ferrooxidans . Appl.Environ.Microbiol.57:2063-2068.(b)

58. Rawling DE (2001) The molecular genetics of Thiobacillus ferrooxidans and other mesophilic, acidophilic, chemolithotrophic, iron or sulfur-oxidizing bacteria. Hydrometallurgy 59:187-201.

59. Rawlings DE (2002) Heavy metal mining using microbes. Ann. Rev. Microbiol. 56: 65-91.

60. Rawlings DE (2004) Microbially assisted dissolution of minerals and its use in the mining industry. Pure Appl.Chem.76(847-856)

61. Romano P, Blazquez ML, Alguacil FJ, Munoz JA, Ballester A, and Gonzalez F. (2001)Comparative study on the selective chalcopyrite bioleaching of a molybdenite concentrate with mesophilic and thermophilic bacteria. FEMS Microbiology Letters. 196:71-75.

62. Sadowski Z, Jazdzyk E, Kavas H. (2003) Bioleaching of copper ore flotation concentrates. Mineral Engineering. 16: 51-53.

63. Santhiya D, Subramanian S, Natarajan KT. (2000) Surface chemical studies on galena and sphalerite in the presence of with reference to mineral beneficiation. Mineral Engineering. 13:747-763.

64. Schnell HA (1997) Bioleaching of copper.72 :21-43.

65. Shankar C, Sridevi D, Joonhong P, Dexilin M, Thamaraiselvi K(2007) Biosorption of chromium and nickel by heavy metal resistant fungal and bacterial isolates. J.Hazard.Mater.146:270-277.

66. Singh AL, (2010) Uptake of Zinc and its localization in different cell sectors of Pseudomonas japonica. World Academy of Science, Engineering and Technology, Penang, Malaysia, 62 :1633-1648.

67. Singh AL, Murali Krishna P, Nageswara Rao T, Sarma P.N (2008) Biosorption of zinc by Staphylococcus aureus from synthetic waste water : Effect of modifying factors. Pollution Res. Environmedia. 27:1-5

68. Singh AL, Sarma P.N (2006).Uptake of Cr (VI) in the presence of sulphate by Bacillus mycoides in aerobic culture. IndianJ.ofBiotechnology, New, Delhi, 5:506-509

69. Singh AL, Singh SK, Rajkumar B. (2000) Removal of Cu by Pseudomonas-aeruginosa from polluted water of Barakar River, Dhanbad, Bihar, India.. Indian Journal of Environmental Geochemistry. 3:7-9.

70. Sung C M, Suk CK, Su KD, Wook RH. (2005) Bioleaching of uranium from low grade black schist by Acidithiobacillus ferrooxidans. World J Microbio. and Biotech. 21:377-380.

71. Tipre DR, Dave S R. (2004) Bioleaching process of Cu-Pb-Zn bulk concentrate at high pulp density. Hydrometallurgy. 75:37-43.

72. Torma AE. (1977) The role of Thiobacillus ferrooxidans in hydrometallurgical processes. In N. Blakebrough, A.Fiechter, and T.K.Ghose(eds.). Advances in Biochemical Enineering . Vol.6.Springer-Verlag, Berlin. 1-37.

73. Valix M, Thangavelu V, Ryan D, Tang J(2009) Using halotolerant Aspergillus foetidus

74. in bioleaching nickel laterite ore. International Journal of Environment and Waste Management. 3:253-264.

75. Van Aswegen PC, Godfrey MW, Miller DM, Haines AK(1991)Developments and innovations in bacterial oxidation of refractory ores. Miner. Melallurg.Process.8:188-192.

76. Wang J, Chen C (2006) Biosorption of heavy metals by Saccharomyces cerevisiae: Rev. Biotechnol.Adv., 24: 427-451.

77. Warhurst A. (1985) Biotechnology for mining: the potential of an emerging technology, the Andean Pact Copper Project and some policy implications. Development and Change, London. 16: 93-121.

78. Watling HR (2006) The bioleaching of sulphide minerals with emphasis on copper sulphids-Areview.Hydrometallurgy 84:81-108.

79. Yuce E A, Tarkan M H, Dogan Z M. (2006) Effect of bacterial conditioning and the flotation of copper ore and concentrate. African Journal of Biotechnology. 55:448-452.

80. Zahoor A, Rehman A (2009) Isolation of Cr (VI) reducing bacteria from industrial effluents and their potential use in bioremediation of chromium containing wastewater. Journal of Environmental Sciences.21: 814-820.

81. Zaied KA, Abd EI- Mageed HN, Fayzalla EA, Sharief AE,Zehry AA (2008) Enhancement Biosorption of heavy metals from Factory Effluents via Recombinants Induced in Yeast and Bacteria. Austr. J. Basic Appl.Sci., 2(3): 701-717.

CHAPTER-8

MICROBIAL INTERACTIONS WITH METAL POLLUTANTS

Prachi Bhargava [1] and Rohit Saluja[2]

1 Amity Institute of Biotechnology, Amity University, Sec-125, Noida, Uttar Pradesh, India.

2 Department of Medicine Solna, Karolinska Institutet, Stockholm, Sweden

ABSTRACT

Metallic elements are ubiquitously present in the environment and living organisms are exposed to heavy metals in nature, commonly present in their ionized species. Metals constitute an important class of toxic substances which exert diverse toxic effects on microorganisms. Metal exposure both selects and maintains microbial variants which are able to tolerate their harmful effects. Varied and efficient metal resistance mechanisms have been identified in diverse species of bacteria, fungi and protists. Microbes interact with metals and minerals in natural and synthetic environments, altering their physical and chemical state, with metals and minerals also able to affect microbial growth, activity and survival. The study of the interactions between microorganisms and metals may be helpful to understand the relations of toxic metals with higher organisms such as mammals and plants. Some microbial systems of metal tolerance have the potential to be used in biotechnological processes, such as the bioremediation of environmental metal pollution or the recovery of valuable metals. This chapter seeks to outline some of the main ways that microbes interact with metals and minerals, their importance in environmental processes, and their applied significance.

INDEX

LAP LAMBERT ACADEMIC PUBLISHING AG & CO. KG, DUDWELLER LANDSTR, GERMANY

Introduction

Metal–mineral–microbe interactions are of key importance within the framework of geomicrobiology and also fundamental to microbial biomineralization processes. The term biomineralization represents the collective processes by which organisms form minerals [1,2], a phenomenon widespread in biology and mediated by bacteria, protists, fungi, plants and animals. Most biominerals are calcium carbonates, silicates and iron oxides or sulfides [1,3]. Biomineralization is itself an important interdisciplinary research area, and one that overlaps with geomicrobiology [2,4,5].

Metal toxicity and biomagnification

Biomagnification refers to the increase in the concentration of certain substances such as pesticides or heavy metals that occurs in a food chain, because of their low rate of degradation, persistancy or food chain energetics.The major sources of toxic metals are organic wastes, industrial effluents, refuse burning, transportation, power generation etc.

On the other hand bioaccumulants are substances that increase in concentration among living organisms as they take in contaminated air, water or food since these substances are metabolized or excreted very slowly-e.g. polychlorinated biphenyl. Heavy metal contamination due to natural and anthropogenic sources is a global environmental concern. Release of heavy metal without proper treatment poses a significant threat to public health because of its persistence, biomagnification and accumulation in food chain. Non-biodegradability and sludge production are the two major constraints of metal treatment [6]. Of studies on metals only organic mercury shows biomagnification and most metals are regulated and excreted and do not biomagnify. Studies on organic compounds 67% claimed to show biomagnification. However, bioconcentration (uptake from the surrounding water) is the most usual way that organic compounds are accumulated in

LAP LAMBERT ACADEMIC PUBLISHING AG & CO. KG, DUDWELLER LANDSTR, GERMANY

organisms from invertebrates to and including fish. Only in sea-birds and marine mammals is food intake the major route and where biomagnification can be clearly shown. Body concentrations of organic compounds vary with lipid content and thus in order to compare across species normalisation to uniform lipid content should be done [7].

The Japanese mass poisoning in 1954's, classically named today as "Minamata disease" in which mercury and its derivatives occupy a special place as the most important pollutants for the human health [8]. This disease occurred in the Minamata Bay of Kyushu (Japan) during 1953-1961 and at Niigata (Japan) in 1965. The deadly tragedy was due to the consumption of heavily mercury contaminated fishes by the villagers which had mercury levels as high as 27-102 ppm i.e on an average 50 ppm. Mercury readily penetrates the Central Nervous System and methyl mercury penetrates through placenta to the unborn foetus. In Minamata, a series of infants poisoned in their mother's womb was recorded by the first investigative team from Kumamoto University [9].The health risks resulting from methylmercury exposure began with ataxia, dysarthria, constriction of visual fields, impaired hearing, and sensory disturbance as symptoms of fatal methylmercury poisoning [10]. Not only humans but also in experimental model systems like drosophila, methyl mercury (0.25 ppm) treatment has been shown to cause chromosomal dysjunction in gametes. Organisms are able to control metal concentrations in certain tissues of their body. This ensures to minimize damage of reactive forms of essential and nonessential metals and to control selective utilization of essential metals [11]. An interesting and also larming observation by Werlin et.al. [12] showed that bare CdSe quantum dots that have accumulated in Pseudomonas aeruginosa bacteria can be transferred to and biomagnified in the Tetrahymena thermophila protozoa that prey on the bacteria. Cadmium concentrations in the protozoa predator were approximately five times higher than their bacterial prey. Quantum-dot-treated bacteria were differentially toxic to the protozoa, in that they inhibited their own digestion in the protozoan food vacuoles. Since the protozoa did not lyse, largely intact quantum dots remain available to higher trophic levels. The observed biomagnification from bacterial prey is significant because bacteria are at the base of environmental food webs.

Effect of metal toxicity on microbes

Metals play an integral role in the life processes of microorganisms. Some metals are serve as micronutrients, are used in redox reactions, help in building the osmotic pressure eg milikecalcium, cobalt, iron, magnesium, manganese, sodium, nickel, zinc, etc [13]. Whereas, there are some heavy metals which do not play any useful roles in the biological processes, are non essential and are toxic to the microorganisms like aluminium, cadmium, lead, mercury, silver, gold, etc.The heavy metals after entering inside the biological system interfere with the normal metabolism of the body and cell. Toxicity occurs through the displacement of essential metals from their native binding sites or through ligand interactions [13,14]. Table 1 shows some of the metals and their harmful effects on metabolism.

Table 1. Metals and their target metabolic activities

S.No	Affected function	Metals involved
1	Cell membrane disruption	Mercury, lead, zinc, nickel, copper, cadmium
2	Inhibition of transcription	Mercury
3	Inhibition of translation	Mercury, lead, cadmium
4	DNA damage	Mercury, lead, cadmium, arsenic
5	Inhibition of enzyme activity (antimetabolite)	Mercury, lead, arsenic, cadmium, copper
6	Protein denaturation	Mercury, lead, cadmium
7	Inhibition of cell division	Lead, cadmium, mercury, nickel

Mechanisms of Microbial Metal Resistance and Detoxification

Recent advances have been made in understanding metal--microbe interaction and their application for metal accumulation/detoxification. Some microbial systems of metal

LAP LAMBERT ACADEMIC PUBLISHING AG & CO. KG, DUDWELLER LANDSTR, GERMANY

tolerance can potentially be used in biotechnological processes, such as the bioremediation of environmental metal pollution or the recovery of valuable metals [15]. Microbial metal bioremediation is an efficient strategy due to its low cost, high efficiency and ecofriendly nature.

Resistance may be general or specific against some metals. It can be plasmid coded or an innate property of the organism. It may or may not be metal dependent. Several -OH scavengers including mannitol and thiourea have a weak capacity to bind metals and withdraw catalytic iron ions from sensitive targets.Binding of metals to extracellular materials immobilizes the metal and prevents its entry into the cell. Metal binding to anionic functional groups on cell surfaces occurs with a large number of cationic metals including Cd, Zn, Pb and Fe. Certain groups like sulfhydryl, carboxyl, hydroxyl, amine, amide, sulfonate, etc which are present on algal surfaces strongly bind metals. Extracellular binding is usually to slime layers or exopolymers composed of carbohydrates, polysaccarides, Extracellular polymeric substances (EPS) are very common in the environment. Extracellular polymeric substances (EPS) play an important role in cell aggregation, cell adhesion, and biofilm formation, and protect cells from a hostile environment.

Another method which microbes have evolved to resist metals are siderophores. Siderophores are produced by bacteria, fungi and graminaceous plants for scavenging iron from the environment. They are low-molecular-weight (500-1500 daltons) substances possessing a high affinity for iron(III) (Kf > 1030), the biosynthesis of which is regulated by iron levels and the function of which is to supply iron to the cell [16]. The main function of siderophores is to make iron content properly available to the microbes in an iron limiting environment. In bacteria, Fe^{2+}-dependent repressors bind to DNA upstream to genes involved in siderophore in high intracellular iron concentrations. At low concentrations, Fe^{2+} dissociates from the repressor, which in turn dissociates from the DNA, leading to transcription of genes. In gram-negative and AT-rich gram-positive bacteria, this is usually regulated by the Fur (ferric uptake regulator) repressor, whilst in

GC-rich gram-positive bacteria (e.g. Actinobacteria) it is DtxR (diphtheria toxin repressor), so-called as the production of the dangerous diphtheria toxin by Corynebacterium diphtheriae is also regulated by this system [17]. These siderophores are excreted into the extracellular environment, where they act to sequester and solubilize the iron. Siderophores are then recognized by cell specific receptors on the outer membrane of the cell [18,19]. Siderophores are also useful in medicine for iron and aluminum overload therapy and antibiotics for better targeting [20]. Among the plant kingdom grasses of Poaceae family are able to efficiently sequester iron by releasing phytosiderophores via their root into the surrounding soil rhizosphere. Some metals other than iron which are chelated by the siderophores are- Aluminum, Gallium, Chromium, Copper, Zinc, Lead, Manganese, Cadmium, Vanadium, Indium, Plutonium, Uranium, etc.

Metallothionein (MT) represents another mechanism by which prokaryotes and eukaryotes protect themselves against metal toxicity. Liver and kidneys synthesize MT In the human body. MT is a low molecular weight protein which belongs to cysteine family. MTs have the capacity to bind both physiological (such as zinc, copper, selenium) and xenobiotic (such as cadmium, mercury, silver, arsenic) heavy metals through the thiol group of its cysteine residues, which represents nearly 30% of its amino acidic residues [21]. Cysteine is a sulfur-containing amino acid, hence the name "-thionein". MTs function is not clear, but experimental data suggest MTs may provide protection against metal toxicity, be involved in regulation of physiological metals (Zn and Cu) and provide protection against oxidative stress. They play an important role in the uptake, transport, and regulation of zinc in biological systems. However, there are data which suggests that some MT forms bind with inorganic sulfide and chloride ions. In some MTs, mostly bacterial, histidine participates in zinc binding. By binding and releasing zinc, metallothioneins (MTs) may regulate zinc levels within the body. When zinc enters a cell, it can bind to thionein ("metallothionein") and can be carried to another part of the cell where it is released to another organelle or protein. In this way the thionein-metallothionein becomes a key component of the zinc signaling system in cells. This system is particularly important in the brain, where zinc signaling is prominent both between and within nerve cells.

Detrimental Impacts of Metal-Microbe Interaction

Microbes encounter metal and metalloids of different kinds in the environment and they have both beneficial and detrimental aspects. Heavy metal contaminants in the environment are eventually deposited in the soil. Mineral rock weathering and anthropogenic sources are the main types of metal inputs to soil. These are some anthropogenic methods by which heavy metals are mixed with soil.

- Metalliferous mining and smelting (arsenic, cadmium, lead and mercury)
- Industry (arsenic, cadmium, chromium, cobalt, copper, mercury, nickel, zinc)
- Atmospheric deposition (arsenic, chromium, cadmium, copper, lead ,mercury, uranium)
- Agriculture (arsenic, cadmium, copper, lead, selenium, uranium, zinc)
- Waste disposal (arsenic, cadmium, copper, lead, chromium, mercury and zinc).

Human diseases have resulted from consumption of cadmium contaminated food. [22, 23]. Studies have shown that long-term heavy metal contamination of soils has harmful effects on soil microbial activity, especially microbial respiration [24]. Aside from long-term metal-mediated changes in soil enzyme activities, many reports have shown large reductions in microbial activity due to short-term exposure to toxic metals [25]. Bacterial activity, measured by thymidine incorporation technique, had been shown to be very sensitive to metal pollution both under laboratory and field conditions [26,27]. Moreover, habitats that have high levels of metal contamination show lower numbers of microbes than uncontaminated habitats [28]. Besides the direct toxic effects of heavy-metal and, in particular, radioactive elements on soil microbiota [29] and plants [30], their mobilisation (solubilisation) and leaching from disposal sites presents a significant environmental threat. Moreover, metals when present in the atmosphere also pose a serious threat to the environment. Acidification of rain-water is identified as one of the most serious environmental problems of transboundary nature. Acid rain is a common term for rain, snow, fog and dew that has a higher than normal acidity. In extreme cases the acidic level of acid rain may reach a pH of 3.2-nearly as acidic as a carbonated beverage. They are a mixture of sulphuric and nitric acids depending upon the relative quantities of oxides of sulphur and nitrogen emissions. Whenever these acids interact with other constituents of the atmosphere, protons are released causing increase in the soil acidity Lowering of soil pH mobilizes and leaches away nutrient cations and increases availability of toxic heavy metals. Acidification of water bodies causes detrimental impact on aquatic organisms including fishes. Acid rain also damages man-made materials and structures.

Bio corrosion

Corrosion causes dramatic economic loss. Corrosion or corrode is derived from the Latin corrodere, which means "to gnaw to pieces." The general definition of corrode is to eat into or wear away gradually. Corrosion is the disintegration of metal into its constituent atoms due to chemical reactions; this means electrochemical oxidation of metals in reaction with an oxidant such as oxygen. Biofilms influence electrochemical processes at the metal surface, often leading to deterioration of metals referred to as biocorrosion or microbiologically influenced corrosion (MIC) [31]. Biofilms typically consist of microbial cells and their metabolic products including extracellular polymers, and inorganic precipitates. Interaction of biofilms and exopolymers with metal ions has long been proposed as one of the mechanisms of metal biodeterioration. Majority of the studies on MIC have concentrated on aerobic/anaerobic bacteria. Both aerobic bacteria, which flourish in an oxygenated environment, and anaerobic bacteria, which thrive in a minimal or non-oxygen environment have been documented in MIC [32]. In some cases, these two bacterial types share a symbiotic relationship as aerobic bacteria deposit biofilms under which oxygen depleted zone is formed at the metal interface. This oxygen depleted zone then becomes an ideal environment for the growth of anaerobic bacteria colonies. Desulfovibrio vulgaris, a sulfur reducing bacteria, generate hydrogen sulfide as a metabolic bi-product. This species has been implicated in MIC in iron, steel, stainless steel, aluminum, zinc and copper alloys

Beneficial Metal-Microbe Interactions

The metal accumulating capacity of microorganisms can be exploited to remove, concentrate and recover metals from mine tailings and industrial effluents [33].While metals which cannot be broken down into non-toxic components like organic compounds, bioremediation can be used to stabilize, extract, or reduce the toxicity of soil and groundwater contaminated by acid mine drainage [34]. On the other hand, they can directly or indirectly affect plant-microbe interactions in the rhizosphere. The rhizosphere, as compared to the bulk soil, is highly populated by various microorganisms mainly comprising bacteria (predominatingly Gram-negative bacteria) and mycorrhizal fungi showing higher metabolic activity [35] even in polluted soils.

Fungal symbiosis in mineral transformation

The plant-associated rhizobacteria and mycorrhizae may significantly increase the bioavailability of various heavy metal ions for their uptake by plants. Also, they are known to catalyse redox transformations leading to changes in heavy-metal bioavailability [36]. Ectomycorrhizal fungi (Suillus granulatus and Paxillus involutus) can release elements from apatite and wood ash (K, Ca, Ti, Mn, Pb) and accumulate them in the mycelia [37]. Ericoid mycorrhizal and ectomycorrhizal fungi can dissolve a variety of cadmium, copper, zinc and lead-bearing minerals, including metal phosphates [38]. Lichens are fungi that exist in facultative or obligate symbioses with one or more photosynthesizing partners; they play an important role in many biogeochemical processes. Lichens are pioneer colonizers of fresh rock outcrops. Lichens can accumulate metals such as lead, copper, and many other elements of environmental concern, including radionuclides, to high levels [39]. They can also form a variety of metal-organic biominerals, e.g. oxalates, especially during growth on metal-rich substrates. For example, on copper-sulfide-bearing rocks, precipitation of copper oxalate (moolooite) can occur within the lichen thallus. Other than the above benefits, microbially mediated bioleaching and biotransformation of heavy metals and radionuclides entrapped within soil minerals are some of the most important useful applications.

LAP LAMBERT ACADEMIC PUBLISHING AG & CO. KG, DUDWELLER LANDSTR, GERMANY

Bioleaching

Bioleaching is a phenomenon in which microorganisms aid in the dissolution of metals from their mineral source. This is one of the most beneficial applications of biohydrometallurgy used to recover copper, zinc, lead, arsenic, antimony, nickel, molybdenum, gold, and cobalt. Commercially available species of Thiobacillus, Leptospirillum, Sulfobacillus, and Sulfolobus are some of the important bacteria used in bioleaching. Different fungal species are also beneficial in this process as they can grow on a variety of substrates, such as electronic scrap, catalytic converters, and fly ash from municipal waste incineration. Two fungal strains (Aspergillus niger, Penicillium simplicissimum) are able to mobilize Cu and Sn by 65%, and Al, Ni, Pb, and Zn by more than 95%.. Aspergillus niger produces citric acid and this form of leaching does not rely on microbial oxidation of metal, but rather uses microbial metabolism as source of acids which directly dissolve the metal.

Enhanced recovery of Petroleum product

Petroleum and oil is a complex mixture of hydrocarbons and other organic compounds, including some organometallo constituents, that plays an important role in providing energy supply to the world. Microbes play a significant role in oil recovery - microbial enhanced oil recovery, is commonly referred to as MEOR (Microbial Enhanced Oil Recovery). There are several traditions in which microorganisms can enhance oil recovery, for example, microorganisms can be used to reduce the paraffin build-up in producing wells or they can be utilized to produce solvents or polymers above ground for pumping into the oil-bearing formation [40]. Two different bacterial strains: P. aeruginosa and P. aeruginosa have been shown the highest producer of biosurfactant that enhanced mineral oil recovery [41]. This technology (MEOR) requires consideration of the physicochemical properties of the reservoir in terms of salinity, pH, temperature, pressure, and nutrient availability [42]. Bacteria are considered promising candidates for microbial enhanced oil recovery. Molds, yeasts, algae, and protozoa are not suitable due to their size or inability to grow under the conditions present in reservoirs. Different microorganism including bacteria, yeasts, fungi, and algae are also considered as of hydrocarbon-degrading organisms [43,44] and responsible for hydrocarbon transformations. Magot et al. [45] have shown the significance of different microorganisms from petroleum reservoirs, including mesophilic and thermophilic sulfate-reducing bacteria, methanogens, mesophilic and thermophilic fermentative bacteria, and iron-reducing bacteria. MEOR currently represent efficient and predictable oil recovery but progress in this area is slow. Development of a universal additive mixture, consisting of a combination of microbial strains, nutrients, surfactants, and buffering agents in appropriate proportions, may represent a further productive line of research.

Conclusion

Microorganisms can change the oxidation state of metals and concomitantly deposit metal oxides and zerovalent metals on or into their cells. The microbial mechanisms involved in these processes have been extensively studied in natural environments and applications of microbe-metal interactions have gained much weight age in biotechnology. Metabolic processes of microorganisms are the cause for the dissolution of minerals, and especially pyrite oxidation results in the generation of acid mine drainage which, in turn, leads to heavy metal contamination as a result of mining activities. On the other hand, microbial metabolism can also contribute to the formation of certain ore deposits over geological time. The adaptation to heavy metal rich environments is resulting in microorgansims which show activities for biosorption, bioprecipitation, extracellular sequestration, transport mechanisms, and/or chelation. Such resistance mechanisms are the basis for the use of microorganisms in bioremediation approaches. The strongly interdisciplinary field of bio-geo-interactions spanning from the microorganism to the mineral make geomicrobiology one of the most important concepts within microbiology holds much promise for future developments in both basic research as well as applied sciences.

REFERENCES:

1. Bazylinski, D. A. (2001). Bacterial mineralization. In Encyclopedia of Materials: Science and Technology, pp. 441–448. Amsterdam: Elsevier.

2. Dove, P. M., De Yoreo, J. J. & Weiner, S. (2003). Biomineralization. Reviews in Mineralogy and Geochemistry, vol. 54. Washington, DC: Mineralogical Society of America.

3. Baeuerlein, E. (2000). Biomineralization. Weinheim, Germany: Wiley-VCH

4. Banfield, J. F. & Nealson, K. H. (1997). Geomicrobiology: Interactions between Microbes and Minerals, Reviews in Mineralogy and Geochemistry, vol. 35. Washington, DC: Mineralogical Society of America.

5. Banfield, J. F., Cervini-Silva, J. & Nealson, K. H. (2005). Molecular Geomicrobiology, Reviews in Mineralogy and Geochemistry, vol. 59. Washington, DC: Mineralogical Society of America.

6. Rajendran, P., Muthukrishnan, J., Gunasekaran, P. (2003) Microbes in heavy metal remediation. Indian J Exp Bio, 41, 935-44.

7. Gray, J.S. (2002). Biomagnification in marine systems: the perspective of an ecologist. Mar Pollut Bull, 45, 46-52.

8. Gras, G. (1981). Mercury and methylmercury pollution of fishery products. Toxicological effects on human health. Mondain J.Toxicol Eur Res,3, 243-59.

9. Harada, M. (1995).Minamata disease: methylmercury poisoning in Japan caused by environmental pollution. Crit. Rev. Toxicol,25, 1-24.

10. Grandjean, P., Hiroshi, S., Katsuyuki, M., Komyo, E. (2010). Adverse Effects of Methylmercury: Environmental Health Research Implications. Environmental Health Perspectives,118 , 1137-45

11. Vijver, M.G., Van Gestel, C.A., Lanno, R.P., Van Straalen, N.M., Peijnenburg, W.J. (2004). Internal metal sequestration and its ecotoxicological relevance: a review. Environ Sci Technol., 38, 4705-12.

12. Werlin, R., Priester, J.H., Mielke, R.E., Krämer, S., Jackson, S., Stoimenov, P.K., Stucky, G.D., Cherr, G.N., Orias, E., Holden, P.A. (2011) Biomagnification of cadmium selenide quantum dots in a simple experimental microbial food chain. Nat Nanotechnol,1, 65-71.

13. Bruins, M.R., Kapil, S., Oehme, F.W. 2000. Microbial resistance to metals in the environment. Ecotoxicol and Environ Safety, 45, 198-207.

14. Nies, D.H. (1999). Microbial heavy-metal resistance. Applied microbiol Biotechnol, 51, 730-750.

15. Cervantes, C., Espino-Saldaña, A.E., Acevedo-Aguilar, F., León-Rodriguez, I.L., Rivera-Cano, M.E., Avila-Rodríguez, M., Wróbel-Kaczmarczyk, K., Wróbel-Zasada, K., Gutiérrez-Corona, J.F., Rodríguez-Zavala, J.S., Moreno-Sánchez, R.(2006) Microbial interactions with heavy metals. Rev Latinoam Microbiol., 48,203-10.

16. Hider, R.C. and Kong, X. (2010) Chemistry and biology of siderophores. Nat Prod Rep., 27,637-57.

17. Miethke, M. and Marahiel, M. (2007). Siderophore-Based Iron Acquistion and Pathogen Control. Microbiology and Molecular Biology Reviews ,71 , 413–451.

18. Kraemer, Stephan, M., Crowley, David, Kretzschmar, Ruben (2006). Siderophores in Plant Iron Acquisition: Geochemical Aspects. Advances in Agronomy, 91: 1–46.

19. Kraemer, Stephan M., Butler, Allison, Borer, Paul, and Cervini-Silva, Javiera (2005). Siderophores and the dissolution of iron bearing minerals in marine systems. Reviews in Mineralogy and Geochemistry, 59, 53–76.

20. del Olmo, A., Caramelo, C., SanJose, C. (2003). Fluorescent complex of pyoverdin with aluminum. Journal of Inorganic Biochemistry , 97, 384–387.

21. Sigel, A., Sigel, H., Sigel, R.K.O., ed (2009). Metallothioneins and Related Chelators. Metal Ions in Life Sciences. 5. Cambridge: RSC Publishing

22. Kobayashi, J. (1978). Pollution by cadmium and the itai-itai disease in Japan. In: Toxicity of Heavy Metals in the Environment. Oehme, F. W. (ed.) Marcel Dekker Inc.: New York, pp.199-260.

23. Nogawa, K., Honda, R., Kido, T., Tsuritani, I., Yamanda, Y. (1987). Limits to protect people eating cadmium in rice, based on epidemiological studies. Trace Subst. Environ. Health, 21, 431-439.

24. Doelman, P.and Haanstra, L. (1984). Short-term and long-term effects of Cd, Cr, Cu, Ni, Pb, and Zn on microbial respiration in relation to abiotic soil factors. Plant Soi., 79, 317-321.

25. Hemida, S. K., Omar, S. A., Abdel-Mallek, A. Y. (1997). Microbial populations and enzyme activity in soil treated with heavy metals. Water Air Soil Pollut, 95, 13-22

26. Diaz-Ravina, M. and Baath, E.(1996). Development of metal tolerance in soil bacterial communities exposed to experimentally increase metal levels. Appl. Environ. Microbiol. 62, 2970-2977.

27. Diaz-Ravina, M. and Baath, E. (1996). Thymidine and leucine incorporation into bacteria experimentally contaminated with heavy metals. Appl. Soil. Ecol, 3, 225-234.

28. Kandeler, E., Tscherko, D., Bruce, K. D., Stemmer, M., Hobbs, P. J., Bardgett, R. D., Amelung, W.(2000). Structure and function of the soil microbial community in microhabitats of a heavy metal polluted soil. Biol. Fertil. Soils, 32, 390-400.

29. Ruggiero, C.E., Boukhalfa, H., Forsythe, J.H., Lack, J.G., Hersman, L.E., Neu, M.P. 2005. Actinide and metal toxicity to prospective bioremediation bacteria Environ. Microbiol., 7: 88-97.

30. Peng, H., Geng, W., Yong-quan, W., Mao-teng, L., Jun, X., Long-jiang, Y.,Effect of heavy metal stress on emerging plants community constructions in wetland. Water Sci Technol. 2010, 62, 2459-66.

31. Zuo, R. (2007). Biofilms: strategies for metal corrosion inhibition employing microorganisms. Appl Microbiol Biotechnol., 76, 1245-53.

32. Hang T. Dinh, Jan Kuever, Marc Mußmann, Achim W. Hassel, Martin Stratmann , Friedrich Widdel. (2004). Iron corrosion by novel anaerobic microorganisms Nature, 427, 829-832.

33. Malekzadeh, F, Farazmand, A., Ghafourian, H., Shahamat, M.,. Levin, M., Colwell, R.R. (2002). Uranium accumulation by a bacterium isolated   from electroplating effluent. World Journal of Microbiology and Biotechnology, 18, 295-302.

34. White, C., Wilkinson, S.C., Gadd, G.M (1995). The Role of Microorganisms in Biosorption of Toxic Metals and Radionuclides. International Biodeterioration and Biodegradation, 35: 17-40.

35. Lugtenberg, B.J.J., Chin-A-Woeng, T.F.C., Bloemberg, G.V. (2002). Microbe-plant interactions: principles and mechanisms. Ant. van Leeuwenhoek, 81: 373.

36. Yang, X., Feng, Y., He, Z., Stoffella, P.J.. (2005). Molecular mechanisms of heavy metal hyperaccumulation and phytoremediation. J. Trace Elem. Med. Biol., 18, 339.

37. Wallander, H., Mahmood, S., Hagerberg, D., Johansson, L., Pallon, J. (2003). Elemental composition of ectomycorrhizal mycelia identified by PCR-RFLP analysis and grown in contact with apatite or wood ash in forest soil. FEMS Microbiol Ecol, 44, 57–65.

38. Leyval, C. and Joner, E. J. (2001). Bioavailability of heavy metals in the mycorrhizosphere. In Trace Elements in the Rhizosphere, pp. 165–185. Edited by G. R. Gobran, W. W. Wenzel & E. Lombi. Boca Raton, FL: CRC Press.

39. Purvis, O. W. and Pawlik-Skowronska, B. (2008). Lichens and metals. In Stress in Yeasts and Filamentous Fungi, pp. 175–200. Amsterdam: Elsevier.

40. Brown, L.R. (2010). Microbial enhanced oil recovery (MEOR). Curr Opin Microbiol., 13, 316-20.

41. Bordoloi, N.K. and Konwar, B.K. (2008). Microbial surfactant-enhanced mineral oil recovery under laboratory conditions. Colloids Surf B Biointerfaces, 63:73-82.

42. Khire, J. M., and Khan, M. I. (1994). Microbially enhanced oil recovery (MEOR). Part 1. Importance and mechanisms of microbial enhanced oil recovery. Enzyme Microb. Technol, 16,170-172.

43. Atlas, R. M., and Cerniglia, C. E. (1995). Bioremediation of petroleum pollutants: diversity and environmental aspects of hydrocarbon biodegradation. BioScience , 45, 332-338.

44. Leahy, J. G., and. Colwell, R. R. (1990). Microbial degradation of hydrocarbons in the environment. Microbiol. Rev, 54, 305-315.

45. Magot, M., Ollivier, B. Patel, B. K. C. (2000). Microbiology of petroleum reservoirs. Antonie van Leeuwenhoek, 77, 103-116.

LAP LAMBERT ACADEMIC PUBLISHING AG & CO. KG, DUDWELLER LANDSTR, GERMANY

CHAPTER-9

MICROBES IN PRODUCTION AND PROCESSING OF TEXTILES

Neema Papnai and Vivek Bajpai
Faculty of Arts, Science and Commerce, Mody Institute of Technology and Science
Laxmangarh-332311, Sikar, Rajasthan (India)

ABSTRACT

Industrialization is the key to development. However, it is also recognized as a root cause of environment pollution. Environment pollution is a worldwide threat to public health which has given new initiatives to restore the environment for ecological reasons. India being a major producer and exporter of textiles, the problem of environmental pollution from textile industry has a serious dimension. There are many ways which help in environment restoration. Microbes play an important role by giving us a helping hand. They are used in preparation and production of fibers such as flax, hemp, biopol etc., production of textile dyes and also in degradation of color effluent.

This chapter highlights on the various uses of microorganisms in textiles i.e. in the growth of fiber yielding plants, separation of fiber from stalk, production and degradation of textile dyes. Number and kinds of micro-organisms associated with textiles and the biochemical changes such organisms induce has also been outlined.

INDEX

LAP LAMBERT ACADEMIC PUBLISHING AG & CO. KG, DUDWELLER LANDSTR, GERMANY

INTRODUCTION

The Indian Textiles Industry has an overwhelming presence in the economic life of the country and plays a pivotal role through its contribution to industrial output, employment generation, and export earnings of the country (Das, B. R. and Hati, S., and India in Business). Textile production involves various processes which are harmful for the living being. Higher attention to the adverse effect of synthetic processing and products on health and environment is required and this call for eco-friendly production and processing.

Microbes are ubiquitous in nature. They inhabit polluted environments and are armed with various resistant and catabolic potentials. The catalytic potential of microbes is enormous and is advantageous to mankind for a cleaner and healthier environment. Potential microbes with broad spectrum of activities are taken from their native habitat and are used for better performance. By such studies, the core problem of pollution is attacked tactfully to benefit mankind with a healthy atmosphere (D. Lalithakumari).

Microbes fall into three categories; bacteria, fungi and algae, although only the first two are generally applicable to textiles (Shirley Technology Ltd). There are many stages in textile production and various processes are involved. In each stage and process microbes play an important role. They help in growth of the fiber yielding plant and preparation of fibers through retting process. Microbes also mediate in production and degradation of xenobiotic compounds like dyes and plastics (D.Lalithakumari). Some microbes help in the break down of waste produced from industry and households and are used in the treatment of wastewater by removing organic materials from the water (Waltz, J. J.).

GROWTH OF FIBER YIELDING PLANT

Many textile fibers are derived from plants, including cotton, flax, ramie, jute and hemp. Microbes help in production of unique fibers and in improving yields of existing fibers. Green plants, because of their photosynthetic process form the base of the food chain and thus the beginning of the energy flow through an ecosystem. Microorganisms exert marked influence on the growth of plants and are responsive to the environmental conditions imposed by plant roots. They are recognized not only as a growth substance but also as having the capacity to produce them (Schmidt, E.L.).

Microorganisms are the only important organisms able to assimilate inorganic elements and incorporate them into organic compounds in living tissues, and therefore form a vital link for plant growth. Bacteria and fungi serve as the major link in the cycling process because they decompose organic tissues and release the elements to the soil or water (Rodney, G. Myatt).

Microorganisms are also responsible for building fertile soil for plants to grow in. Microbes stick to the roots of plants and decompose dead organic matter into food for the plant to absorb. Some microorganisms kill the bacteria on plant that make them sick (Library.thinkquest.org).

PRODUCTION OF FIBERS AND BIOPOLYMERS

Natural textile fibers such as cotton, wool and silk are an asset. Different biotechnological routes for cellulose production are being worked out globally. Cellulose is produced as an extra cellular polysaccharide by several bacteria in the form of ribbon-like micro fibrils, and can be used to produce materials of relatively high strength (www.biotechcareer.org). Culturing cells of various strains of Gossypium can produce cotton fibers in vitro with desirable properties. Plant tissue culture provides a steady, all year supply of products without climatic or geographic limitations free of contamination from pests.

Fermentation is developing biopolymers at large-scale. Bacterial storage compound polyhydroxybutyrate (PHB) is developed by Zeneca Bio products and is produced as 'Biopol'. It is high molecular weight linear polyester (melts at 180°C) that can melt spun into biocompatible and biodegradable fibers suitable for surgical use where human body enzymes slowly degrade sutures.

Polysaccharides chitin, alginate, dextran and hyaluronic acid biopolymers (fibers) are used in wound healing as chitin and its derivative chitosan. Chitin and chitosan are an important component of fungal cell wall at present manufactured from sea food (shellfish) wastes. Researches have been directed for use of intact fungal filaments as a direct source of chitin or chitosan fiber to produce inexpensive wound dressings. A wound dressing based on calcium alginate fibers is developed by Courtaulds and is marketed as 'Sorbsan'. Present supply of this polysaccharide relies on brown seaweed for its extraction. However, a polymer of similar structure can also be produced by fermentation from certain species of bacteria. Dextran which is manufactured by fermentation of sucrose by Leuconostoc mesenteroides or related species of bacteria is also developed as non-woven for specialty end-uses such as wound dressings (www.biotechcareer.org).

Proteins are interesting biopolymers where animal and plant protein genes (e.g. collagen, various silks) are transferred into suitable microbial hosts which produce proteins by fermentation.

SEPARATION OF FIBER FROM PLANT

Natural cellulosic fibers are elongated cells abundant in the bark and leaves of some plants. Flax is the most important bast fiber. Others are ramie, jute, sisal, hemp, etc. Lignin is a complex chemical compound most commonly derived from wood and is an integral part of the cell wall of plants. It is the most abundant organic polymer on earth after cellulose, employing 30% of non-fossil organic carbon and constituting of a quarter to a third of the dry mass of wood. A process called retting is employed to separate fibers of commercial importance from the stem of the plants. The fibers are separated by decomposing the cementing material holding them together. The available retting processes are: mechanical retting (hammering), chemical retting (boiling & applying chemicals), steam/vapor/dew retting, and water or microbial retting. Among them, water or microbial retting is a century old and the most popular process in extracting fine bast fibers. This process involves the action of bacteria and moisture to dissolve and rot away the cellular tissues and gummy substances that surround the fiber bundles in the plant. The bacteria used for the purpose are Clostridium butyricum and C. felsinum. The organisms gradually develop and multiply by utilizing free sugars, pectins, hemicellulose and protein of the plants as nutrients (Majumdar and Day; Mohiuddin.et.al., and Sarkar). Among the fungi, Aspergillus nifer,

Macrophomine phaseolina, Chaelomium sp., Phoma sp. (Ahamed, 1963) and several Penicillium sp. has been found to be good retting agents. Several aerobic bacteria (Jalaluddin, 1965) of the genus Bacillus, viz., B. subtilis, B. polymyxa, B. mesentericus, B. macraraus (Bhattcharyya) and anaerobic bacteria of genus clostridium viz., C. tertium, C .aurantibutyricum, C. felsineum etc (Alam) have been isolated for water retting. The microbes which are residents of retting water penetrate into the cortex and phloem regions of the bark through openings in the epidermis and cut ends of the stems, and attack the interconnecting tissues between the filaments. As the retting proceeds, the undesired tissues and intercellular binding substances are gradually degraded into water soluble organic compounds and are consumed by the microbes. Once the surrounding tissue and other substances are dissolved, they fall away. Fiber are then washed with water and separated from the the woody core (the xylem) and sometimes from the epidermis as well. The fibers are finally dried in the sun (Pandey, R). Both bacteria and fungi can decompose cellulose but are dependant on moisture content.

During retting, microbial activity causes a partial degradation of the components that binds tissue together, thereby separating the cellulosic fibers from non fiber tissue. This technique is usually dependent on weather and water resources. Normally, stored premature wood is rapidly attacked by micro organisms, mainly fungi. White rot fungi (Datronia sp. and Oligoporous sp.) belonging to the class Besidimycetes is capable of degrading the woody polymers. The use of fungi not only aims at making production more effective but also environmental friendly. Fungi results in higher pectinolytic activity. The increase in pectinolytic enzyme activity was related to the better efficiency of fiber separation (Tamburini, Elena. et. al.,). A number of bacteria and moulds can digest pectin. This permits the fiber bundles to be separated mechanically from the stems and from each other. Fibers can then be collected and woven into linen or used in the form of ropes and packaging material. The stems are monitored during retting to avoid excessive degradation of the fibers.

LAP LAMBERT ACADEMIC PUBLISHING AG & CO. KG, DUDWELLER LANDSTR, GERMANY

In this way fibers from the stem of jute (Corchorus capsularis and C. olitorius), flax (Linum usitatissimun) and hemp (Cannabis sativa) are retted, and are used in making gunny bags, carpets, ropes, etc.

There are two basic processes of retting. It can be done either with the help of water or with the help of dew (www.microbiologyprocedure.com). In the first process, the cut crop of flax or hemp is spread out on the field/ ground, particularly under somewhat acidic conditions such as would occur in peat moss and is exposed to the air and rain for some time called dew retting. Under these conditions various microorganisms such as moulds and bacteria grow which produce pectinolytic enzyme. Action of pectinolytic enzyme degrades pectin and lignin in middle lamella of plant fibers thus freeing the cellulose fibers. In the second process, the plant stalks are immersed in flowing stream, pond or tank of water and are allowed to remain there for several days. During this period the anaerobic organisms especially Clostridium felsineum digest the pectin. This process, when properly operated yields a nicer fiber which can be made into linen of quite good quality.

The aerobic organisms grow first and consume most of the dissolved oxygen, ultimately creating an environment favorable for the growth of anaerobes. It has been observed that the greater part of decomposition is carried out by anaerobic species. In microbial retting, pectin and hemicellulose are decomposed into water soluble compounds by specific enzymes secreted by the micro-organisms present in water and in plant. Bacterium felsenium considerably shorten the retting period and improve the fiber quality (Ali. et. al.1972).

Jute is a ligno-cellulosic bast fiber. The jute fiber comes from the stem and ribbon (outer skin) of the jute plant. The retting process consists of bundling jute stems together and immersing them in running water. After the retting process, stripping begins. In the stripping process, non-fibrous matter is scraped off and then the fibers are extracted out from within the jute stem (Subramanian, P). Pre-microbial treatment helps jute prevent the development of odor. Resistance of jute to microbial damage may be dependant upon the physical properties (such as, crystalline, chain length, orientation of ultimate cells etc.) and the chemical constituents (such as lignin, hemicelluloses and gummy materials) of the plant. Higher the lignin content, more resistant the fiber will be to rotting i.e. fungal attack. Less rotting will equate to less odor. It is evident that jute extracts show antimicrobial activity by Ilhan et. al.

PRODUCTION OF TEXTILE DYES

The textile dyeing and finishing industries use wide variety of dyestuffs due to the rapid changes in the customers' demands. More than 100,000 commercially available dyes are known and the world annual production of the dyestuffs amounts more than 7×10^5 tones (Robinson **et. al.,**).

Dyes are used as textile colorants designed to form bond with fibers. Some dyes are produced by fermentation. Microorganisms offer great potential for direct production of novel textile dyes or dye intermediates by controlled fermentation techniques replacing chemical synthesis. Many microorganisms produce pigments during their growth which are substantive and indicated by permanent staining. Some species produce up to 30% of their dry weight as pigment. Such microbial pigments are benzoquinone, naphthoquinone, anthraquinone, perinaphthenone and benzofluoranthenequinone derivatives, resembling in some instances the important group of vat dyes (Perumal, K. and Devi, Sagarika).

The global interest and demand in application of the fungal pigments such as carotenoids, flavonoids, betalains, quinones and some tetrapyrroles in dyeing of cotton, silk and wool has been reported in several studies. Production of fungal pigments has increased due to the toxicity problems caused by those of the synthetic origin. Intensive research was undertaken by the MCRC team to culture five fungi namely **Phoma foveata, Curvularia lunata, Alternaria alternata, Sclerotinia** sp. and **Pestalotia** sp. for optimization of the best growth media for maximum mycelia growth and pigmentation (Perumal, K. and Devi, Sagarika).

Many lichens give good dyes. Lichens are unique organisms which combine fungal and algal partners into a single symbiotic relationship. They grow very slowly on rocks, bare soil, tree trunk, hot or cold and wet or dry climates. Several kinds of lichen give yellow, gold and brown color when simmered in a dye pot. Some also impart pleasant fragrance to dyed wool. Other lichens contain pigment called orchils. When treated with ammonia and oxygen develop into rich red, magenta and purple dyes (Buchanan, R).

Researches also focus on extraction, estimation and purification of pigment for development of an eco-friendly product for textile dyeing industry. Modified Czapedox broth amended with dextrose and peptone was found to be the best growth media resulting in good biomass and pigmentation yield for Phoma foveata (64.36 g/L at pH 6.6) and *Pestalotia* sp. (25.32g/L at pH 5.8), while a pH of 6.8 with casein as nitrogen source was favorable for the growth of C. *lunata*. The maximum growth of 13.74g/L (15 th day) in the natural media was recorded in Potato Carrot broth for Alternaria alternata. The pigment extracts of the above fungi were found to be light and temperature stable with good dyeing efficacy using 1-4% alum. Fungal pigment of *Sclerotinia* sp. used on the pre-mordanted cotton yarn showed different shades of light green, green, brown, and pink (Perumal, K. and Devi, Sagarika).

MCRC dye research team is has also been working incessantly on natural dyes extraction from higher plants, mushrooms, bacteria and fungi since two decades. A novel technology involving microbes for the efficient extraction of indigo from Indigofera tinctoria has been developed in the projects entitled "Hyper production of dyes/pigments from selected lower fungi (Curvularia sp., Sclertotinia sp., Pestalotia sp., Phoma sp.)" and "Yield enhancement and pilot scale production of Coriolus versicolor and Ganoderma lucidum and application of its pigments in selected textile dyeing industries". Pigment producing fungi, both microfungi and basidiomycetes mushrooms have been optimized for their extra cellular and intra cellular pigment yield. Extraction, estimation and purification of pigments from selected microfungi was carried out and attempted for development of eco-friendly products for textile dyeing industry. Modified Czapedox broth with peptone as carbon source was found to be the best growth media resulting in good biomass and pigmentation yield for *Phoma foveata* (64.36 g/L) and *Pestalotia* **sp.** (25.32g/L). Bacterial studies have also been successfully undertaken to explore blue bacterial pigments from *Pseudomonas* sp. The organic cultivation of *Ganoderma lucidum* and the pigments is also being explored. The extracted pigments (yellow, black, violet, red and magenta) have been evaluated for their dyeing performance on cotton and silk fabric/yarn (Perumal, K. and Devi, Sagarika).

TEXTILE EFFLUENT TREATMENT

The industrial effluent contains toxic and hazardous pollutants. One particular class of toxin of major concern is synthetic dyes and dye intermediates. Synthetic dyes are widely used in a variety of products, of which textiles are the primary. The worldwide annual textile production is more than 30 million tones with an expected growth of 3% per annum. Dyes and dyestuffs find use in a wide range of industries but are of primary importance to textile manufacturing. Wastewater from the textile industry contains a variety of polluting substances including dyes. It has been estimated that more than 10 - 15 % of the total dyestuff used in dye manufacturing and textile industry is released into the environment during their synthesis and dyeing process (Sarayu and Sandhya).

Almost 2, 80,000 tones of textile dyes are discharged every year worldwide (Mass and Chaudhari). During textile processing, dyeing result in large amounts of the dyestuff being directly lost to the waste water which ultimately finds its way into the environment. The efficient removal of dyes from textile industry effluents is still a major environmental challenge. Environmental legislation is being imposed to control the release of dyes, in particular azo-based compounds into the environment (McMullan et.al.,). Waste streams generated from textile industries are hazardous and difficult to biodegrade owing to the presence of recalcitrant dyes and pigments. The presence of even trace concentration of dyes in effluent is highly visible and undesirable. The release of colored wastewater in the ecosystem is a remarkable source of esthetic pollution, eutrophication, and perturbations in aquatic life (the absorption of light due to textile dyes creates problem to photosynthetic aquatic plants and algae) and cause considerable damage to the receiving water bodies (Sarayu and Sandhya). Often the degradation products of textile dyes are also carcinogenic. Textile industry wastewater due to the presence of dyes is difficult to treat by traditional treatment technology. Some dyestuffs are highly structured polymers and are very difficult to decompose (Sarayu and Sandhya). Microorganisms have the ability to decolorize and metabolize dyes and has attracted interest in the use of bioremediation based technologies for treating textile wastewater. Mechanisms by which diverse categories of microorganisms, such as the white-rot fungi and anaerobic bacterial consortia, bring about the degradation of dyestuffs are being investigated by McMullan et. al.

In textile industry, color removal from dye house effluent, and toxic heavy metal compounds and pentachlorophenol used overseas as a rot-proofing treatment of cotton fabrics poses a challenge for disposal. Microbes or their enzymes are being used to degrade toxic wastes. Pseudomonas aeruginosa decolorizes some di-azo dyes like navitan fast blue at concentrations up to 1200 mg l^{-1} and the organism is able to decolorize various other tannery dyes at different levels. The organism requires ammonium salts and glucose to co-metabolize the dye. Organic nitrogen sources did not support appreciable decolorization whereas, combined with inorganic nitrogen (NH_4NO_3) there was an increased effect on both growth and decolorization. Decolorization of dye starts when the organism reaches late exponential growth phase and after 24 h of incubation nearly 90% of 100 mg l^{-1} of the dye was decolorized (Nachiyar, C.V. and RajKumar, G.S.).

Through experimental watching on textile wastewater, it was observed that textile wastewater treatment under aerobic conditions is possible. Reductive cleavage of the N=N bond is the initial step of the bacterial degradation of azo dyes. Bacterial isolates were isolated by enrichment culture techniques for dye decolorization and degradation from sewage, tannery and pulp and paper mill treatment plants and were bio-chemically characterized and tested for decolorization and biodegradation. The most efficient of them were identified and deposited with IMTECH, Chandigarh and they belong to *Pseudomonas aeruginosa, Bacillus latrosporus* and *Alkaligens* sp. The consortia containing them were used for the wastewater decolorization (Sarayu and Sandhya).

Microbes also play a very important role in the mineralization of pollutants either by natural selection or through recombinant DNA technology making bioremediation process an extension of normal microbial metabolism. Different, pure isolates of *Pseudomonas* sp.

have been well characterized for complete and partial mineralization of morpholine, methyl parathion and other organophosphorous pesticides and fungicides. Other isolates of bacteria, Serratia sp. and Bacillus sp. have also been characterized and documented for their ability to degrade benzimidazole compounds and to effectively decolorize distillery and textile mill effluents respectively (D. Lalithakumari).

Rich collection of bacteria are used which are capable of degrading azo reactive dyes of commercial textile mill like Black B, Turquoise Blue GN, Yellow HEM, Red HEFB and Navy HER along with a number of mono, bi, poly azo dyes and triphenylmethane dyes like Methyl red, Acid black 53, Acid black 76, Acid black 210, Acid green, Acid brown, Sudan black, Sudan IV and Crystal violet. The organisms employed are Serratia marinorubra, Bacillus sp. YW and YDLK consortia for decolorization of textile mill effluents. They are capable of effective decolorization of a wide range of dyes. The biomass concentration of 20g wet weight/L (w/v) and 5 h of treatment time at room temperature with 250 rpm of agitation rate was able to decolorize the textile mill effluents up to a depth of 160cm in static conditions within 5 h with external supplement and nutrition. Bacterial extra cellular polysaccharides (EPS) as biomatrix (bio-reactor) for the decolorization of textile mill effluents and dyes were standardized using pesticides as energy source for enhanced EPS production (D.Lalithakumari).

The effluent of textile mill, distillery, pharmaceutical and tannery were decolorized using both pure culture of bacteria and microbial consortium. Significant reduction in the Biological Oxygen Demand (BOD) and Chemical Oxygen Demand (COD) values of the textile mill effluent by 96% and 94% respectively, was achieved along with 100% decolorization of textile mill effluents. The treatment studies revealed that the organism grew effectively in the raw effluent and further dilution of the effluent resulted in faster decolorization and degradation of dyes (D.Lalithakumari).

Bacillus sp. YW, YDLK consortia and Trichoderma viride are able to bring about 90% color reduction in distillery effluent with an aeration rate of 2kg oxygen / L along with the addition of 1 % DAP as nitrogen source. The colorants and the COD components of the effluent after biological treatment was reduced up to 95% and BOD values reduced up to 96%. Both batch and continuous treatment systems in laboratory and pilot scale experiments were standardized for obtaining the most suitable treatment system for decolorization of the distillery effluent (D.Lalithakumari).

CONCLUSION

Eco-friendly microbes are generally accepted as an environmentally sound and economically feasible protocol for the production of textiles and also for the treatment of hazardous waste and effluents. Researchers has successfully developed new products, opened up new doors, expedited production and helped to clean up environment through the use of microbes in improving fiber yielding plant varieties, production of fibers, improving the fiber properties, production of dyes and pigments and finally in degrading the waste generated from the industry. Undoubtedly, use of microbes can be expected to expand into many other areas of textile industry replacing existing chemical or mechanical processes. Eco-friendly and economical Systems are also developed and could effectively be integrated with physico-chemical methods for pollution control. It is being treated as upcoming science with enormous commercial implications for many industrial sectors in years to come. Innovative interventions for exploring microbes as safer textile dyes are a promising step towards unfurling the technical and commercial viability of bio-pigments as environment friendly textile colorants.

REFERENCES

1. Ahamed, M. (1963). Studies on jute retting bacteria. J. Appl. Bacteriol., 26, 117-126.

2. Alam, S. M. (1970). Jute retting bacteria from certain ditches of east Pakistan. Pak. J. Sci. Indus. Res., 12, 229-231.

3. Ali, M.M., Sayem, A.Z.M and Eshaque, M. (1972). Effect of neutralization of retting liquor on the progress of retting and quality of fiber. Sci. Ind., 7, 134-136.

4.   Bhattacharyya, S.K. (1971-74). Retting of jute key process that needs more attention. Jute bulletin, 194-198.

5.   Biotechnology a boon to Textile Industry. (2010). www.biotechcareer.org/ biotechnology-a-boon-to-textile-industry.

6.   Biotechnology Means Business: state of the art report on 'The Textile & Clothing Industries'. (1995). The Biotechnology Unit, DTI, LGC, Queens Rd., Teddington, Middlesex, TW11 0LY, UK.

7.   Buchanan, Rita. (1999). A weaver's garden: growing plants for natural fibers and dyes.Dyes from plants. New York. Dover publications, inc. 31 east 2nd street, Mineola. ISBN: 0-486-40712-8 1987, 55-123p

8.   D. Lalithakumari. (2005). Microbes: "a tribute" to clean environment. www.envismadrasuniv.org/newsletter1.htm

9.   Das, B. R. and Hati, S. (2009). Indian Textile and Clothing Sector Poised for a Leap. www.fibre2fashion.com

10.   Elena Tamburini, Alicia Gordillo León, Brunella Perito and Giorgio Mastromei. (2003).Characterization of bacterial pectinolytic strains involved in the water retting process. Environmental Microbiology, 5 (9),730-736.

11.   India   in   Business.   http://www.indiainbusiness.nic.in/industry-infrastructure/industrial-sectors/textile.htm

12.   Jalaluddin, M. (1965). Studies on jute retting aerobic bacteria. Econ. Bot., 19, 184-193.

13.   İLHAN, Semra., SAVAROĞLU, Filiz and ÇOLAK, Ferdağ (2007). Antibacterial and Antifungal Activity of Corchorus olitorus L. (Molokhia) Extracts. International journal of Natural and Engineering Sciences, 1(3), 59-61.

14.   Majumdar, A.K. and Day, A. (1977). Chemical constituents of jute ribbon and the materials removed by retting. Food farming and agric., 21, 25-26.

15. Mass, R. and Chaudhari, S. (2005) Adsorption and biological decolorization of azo dye reactive red 2 in semicontinous anaerobic reactor. Process Biochem, 40, 699-705.

16. McMullan, G., Conneely, A., Kirby, N., Robinson, T., Nigam, P., Banat, I. M., Marchant, R. and Smyth, W. F. (2001). Microbial decolorisation and degradation of textile dyes. Applied Microbiology and Biotechnology, 56 (1-2), 81-87.

17. Mohiuddin, G., Chowdhury, M.I., Kabir, A.K.M.R. and Hasib, S.A. (1978). Lignin content of jute cuttings in the white and tossa varities of Bangladesh jute. Bangladesh J. Jute Fib. Res., 3, 27-32.

18. Nachiyar C. Valli and Rajkumar, G. Suseela. (2003). Degradation of a tannery and textile dye, Navitan Fast Blue S5R by Pseudomonas aeruginosa. World Journal of Microbiology and Biotechnology, 19 (6), 609-614.

19. Pandey, Ritu. (2009). Processing of flax fiber. http://agropedia.iitk.ac.in/?q=content/processing-flax-fibre

20. Perumal, K. and Devi, Sagarika. (2010). Eco-friendly Textile Colorants from Microbes http://ifff2010.eu/abstracts.htm

21. Perumal, K. and Devi, Sagarika. (2010). Optimization and production of eco-friendly textile dyes/pigments from Curvularia lunata, Alternaria alternata, Sclerotinia sp., Pestalotia sp., and Phoma foveata. http://ifff2010.eu/abstracts.htm

22. Pilanee, Vaithanomsat., Phusanakom, Poom., Waraporn Apiwatanapiwat and Molnapat Songpim. (2009). A microbiological technique for the separation of Hibiscus sabdariff L. fibers. Journal of Bacteriology Research, 1 (4), 39-45.

23. Robinson, T., McMullan, G., Marchant, R. and Nigam, P. (2001) Remediation of dyes in Textile effluent: a critical review on current treatment technologies with a proposed alternative. Bioresour. Technol, 77, 247 -255.

24. Rodney, G. Myatt. Flora-plant. http://ashvital.freeservers.com/plant.htm

25. Sarayu, K and Sandhya, S. (2009). Potential of facultative micro organisms for bio treatment of textile wastewater. http://www.envismadrasuniv.org. Newsletter, 7 (2).

26. Sarkar, P.B. (1963). The chemistry of jute lignin, Part-II: Potash fusion of lignin. J.Indian chemical soc., 10, 263.

27. Sarkar, P.B. (1964). The chemistry of jute lignin, Part-III: Action of nitric acid on lignin. J.Indian chemical soc., 11, 407.

28. Schmidt, E.L. (1951). Soil microorganisms and plant growth substances: I. historical. Soil Science, 51(2), 129-140.

29. Schmidt, E.L. (1951). Soil Microorganisms and Plant Growth Substances: I. Historical. Soil Science, 71(2), 129-140.

30. Stolz, A. (2001). Basic and applied aspects in the microbial degradation of azo dyes. Appl. Microbiol. Biotechnol, 56, 69 - 80.

31. Subramanian, Pallatheri. Method of using lignin rich fabrics to prevent microbial growth in cleaning materials.   http://www.faqs.org/patents/app/20090208366 Textile microbiology and retting.

32. http://www.microbiologyprocedure.com/industrial-microbiology/textile-microbiology-and-retting.htm

33. The layman's guide antimicrobial fabrics and testing methods. Shirley Technologies Limited. http://home2.btconnect.com/Shirley-Tech/pdf/MICRO-ARTICLE-300404.pdf

34. Waltz, Jay.J. (2010). How Microbes Can Be Helpful or Harmful. www.ehow.com/facts_7323166.

35. www. library.thinkquest.org/CR0212089/microorganisms.

CHAPTER-10

Endophytic bacteria Mediated Phytoremediation

Parmeela[1] and Prabhat Nath Jha[2]

1Department of Microbiology, College of Basic Sciences& Humanities, GB Pant University of Technology& Science, Pantnagar

2. Birla Institute of Technology & Science, Pilani- 333031, Rajasthan

ABSTRACT

Trace of heavy metals such as cadmium (Cd (II)), lead (Pb (II)), copper (Cu (II)) and chromium (Cr (VI)), which are commonly presented in contaminated soils, can enhance Fe deficiency symptoms both in microbes and plants (Baysse et al., 2000; Yoshihara et al., 2006; Christian et al., 2008), thus affecting their growth negatively. Furthermore, they can combine with sulfhydryl groups of proteins, restraining the activity of enzymes. The permanent existences of cadmium (Cd (II)), lead (Pb (II)), copper (Cu (II)) and chromium (Cr (VI)) in polluted ecosystems threaten the health of entire human beings all the time (Nogaw and Kido, 1996). A lot of physicochemical strategies, such as filtration, chemical precipitation, electrochemical treatment, oxidation/reduction, ion exchange, membrane technology, reverse osmosis, and evaporation   recovery, have been developed for removing heavy metals from the polluted water (Xiao et al., 2010). However, most of them appear to be expensive, low efficient, labor-intensive operational or lack of selectivity in the treating process (Chen et al., 2008; Tang et al., 2008).

INDEX

INTRODUCTION

The continued industrialization of countries has led to extensive environmental problems. A wide variety of chemicals/contaminants/ toxic wastes such as heavy metals, persistent organic pollutants, including pesticides, have been detected in different biota such soil, water, and air (i).Technogenic activities (industrial—plastic, textiles, microelectronics, wood preservatives; mining—mine refuse, tailings, smelting; agrochemicals—chemical fertilizers, farmyard manure, pesticides; aerosols— pyrometallurgical land automobile exhausts; biosolids—sewage sludge, domestic waste; fly ash—coal combustion products) are the primary sources of heavy metal contamination and pollution in the environment in addition to geogenic sources. During the last two decades, bioremediation has emerged as a potential tool to clean up the metal-contaminated/ polluted environment. Exclusively derived processes by plants alone (phytoremediation) are time-consuming. Further, high levels of pollutants pose toxicity to the remediating plants. This situation could be ameliorated and accelerated by exploring the partnership of plant–microbe, which would improve the plant growth by facilitating the sequestration of toxic heavy metals. Plants can bioconcentrate (phytoextraction) as well as bioimmobilize or inactivate (phytostabilization) toxic heavy metals through in situ rhizospheric processes. The mobility and bioavailability of heavy metal in the soil, particularly at the rhizosphere where root uptake or exclusion takes place, are critical factors that affect phytoextraction and phytostabilization. Developing new methods for either enhancing (phytoextraction) or reducing the bioavailability of metal contaminants in the rhizosphere (phytostabilization) as well as improving plant establishment, growth, and health could significantly speed up the process of bioremediation techniques. In this review, we have highlighted the role of plant growth promoting rhizo- and/or endophytic bacteria in accelerating phytoremediation derived benefits in extensive tables and elaborate schematic sketches. Continued worldwide industrialization has caused extensive environmental and human health problems. A wide variety of chemicals, e.g., heavy metals, pesticides, chlorinated solvents, etc., have been detected in different natural resources such as soil, water, and air (Mansour and Gad, 2010). Among the pollutants, the heavy metals are of concern to human health due to their cytotoxicity, mutagenicity, carcinogenicity, non biodegradability and Bio-accumulation (Lim and Schoenung, 2010).

Trace of heavy metals such as cadmium (Cd (II)), lead (Pb (II)), copper (Cu (II)) and chromium (Cr (VI)), which are commonly presented in contaminated soils, can enhance Fe deficiency symptoms both in microbes and plants (Baysse et al., 2000; Yoshihara et al., 2006; Christian et al., 2008), thus affecting their growth negatively. Furthermore, they can combine with sulfhydryl groups of proteins, restraining the activity of enzymes. The permanent existences of cadmium (Cd (II)), lead (Pb (II)), copper (Cu (II)) and chromium (Cr (VI)) in polluted ecosystems threaten the health of entire human beings all the time (Nogaw and Kido, 1996). A lot of physicochemical strategies, such as filtration, chemical precipitation, electrochemical treatment, oxidation/reduction, ion exchange, membrane technology, reverse osmosis, and evaporation   recovery, have been developed for removing heavy metals from the polluted water (Xiao et al., 2010). However, most of them appear to be expensive, low efficient, labor-intensive operational or lack of selectivity in the treating process (Chen et al., 2008; Tang et al., 2008).

During the last two decades, bioremediation has emerged as a potential tool to clean up the metal-contaminated/ polluted environment. Exclusively derived processes by plants alone (phytoremediation) are time-consuming. Further, high levels of pollutants pose toxicity to the remediating plants. This situation could be ameliorated and accelerated by exploring the partnership of plant–microbe, which would improve the plant growth by facilitating the sequestration of toxic heavy metals. Plants can bioconcentrate (phytoextraction) as well as bioimmobilize or inactivate (phytostabilization) toxic heavy metals through in situ rhizospheric processes. The mobility and bioavailability of heavy metal in the soil, particularly at the rhizosphere where root uptake or exclusion takes place, are critical factors that affect phytoextraction and phytostabilization. Developing new methods for either enhancing (phytoextraction) or reducing the bioavailability of metal contaminants in the rhizosphere (phytostabilization) as well as improving plant establishment, growth, and health could significantly speed up the process of bioremediation techniques.

Bioremediation is the use of biological agent to remediate contaminats in the environment, Phytoremediation is the use of plants to remediate polluted soils, an eco-friendly and cost effective technology that is currently receiving considerable global attention (Glick, 2010). It is an *in situ,* solar powered remediation technology that requires minimal site disturbance and maintenance resulting in a low cost and a high public acceptance. Since conventional remediation options currently available are frequently expensive and environmentally invasive, phytoremediation turns out to be a valuable alternative, especially for the treatment of large contaminated areas with diffuse pollution. Large-scale applications of phytoremediation still face a number of obstacles, including the levels of contaminants (being toxic for the organisms involved in remediation), the bioavailable fraction of the contaminants (being too low) and, in some cases, evapotranspiration of volatile organic pollutants from soil or groundwater to the atmosphere. A large number of plant species are capable of hyperaccumulating heavy metals in their tissues; however, phytoremediation in practice has several constraints at the level of sites as these are with a variety of different contaminants (Wu et al., 2006a). Further, the success of phytoremediation of metals depends upon a plant's ability to tolerate to accumulate high concentrations of the metals, while yielding a large plant biomass (Grčman et al., 2001). Due to their importance for practical applications, metal-tolerant plant– microbe associations have been the objective of particular attention due to the potential of microorganisms for bioaccumulating metals from polluted environment or its effects on metal mobilization/immobilization and consequently enhancing metal uptake and plant growth (Fig. 1). Synergistic use of plants and microbes has been profitable for cleanup of metalliferous soils (Jing et al., 2007; Glick, 2010).

Endophytic bacteria can be defined as bacteria colonizing the internal tissues of plants without showing symptoms of infection or negative effects on their host(Holliday, 1989; Schulz & Boyle, 2006), and of the nearly 300 000 plant species that exist on the earth, each individual plant is host to one or more endophytes (Strobel et al., 2004). The association of endophytes with their host is varied and complex and we are only staring to understand these interactions. Bacteria and fungal symbionts exist across all areas of life. The use of plants and endophytic bacteria to clean up environmental pollutants has gained momentum in past years. Owing to its green approach, it is gaining significant amount of public attention. Recently, work has been done showing that plant endophytes might be partially responsible for the degradation of environmental toxins. Work at the University of Iowa has shown that a newly discovered organism Methylobacterium populum sp. nov., strain BJ001 ((Van Aken et al., 2004a), which exists as a plant endophyte, is involved in the degradation of energetic compounds such as 2,4,6-trinitrotoluene (TNT), hexahydro-1,3,5-trinitro- 1,3,5-triazine (HMand hexahydro-1,3,5- trinitro-1,3,5- triazine (RDX)(Van Aken et al., 2004b). Other work has suggested that not only do plant endophytes have a role in toxin degradation, but also that the presence of some toxins can affect the make-up of endophytic populations. Siciliano et al. (Siciliano et al., 2001) showed that the genes encoding catabolic pathways increased within the root endophyte population in response to the presence of a given pollutant. Endophytic bacteria have been isolated from many different plant species (Barzanti et al., 2007; Mastretta et al., 2009); in some cases, they may confer to the plant higher tolerance to heavy metal stress and may stimulate host plant growth through several mechanisms including biological control, induction of systemic resistance in plants to pathogens, nitrogen fixation, production of growth regulators, and enhancement of mineral nutrients and water uptake (Ryan et al., 2008). Additionally observed beneficial effects due to bacterial endophytes inoculation are plant physiological changes including accumulation of osmolytes and osmotic adjustment, stomatal regulation, reduced membrane potentials, as well as changes in phospholipid content in the cell membranes (Compant et al., 2005).

Further, the endophytic bacteria isolated from metal hyperaccumulating plants exhibit tolerance to high metal concentrations (Idris et al., 2004). This may be due to the presence of high concentration of heavy metals in hyperaccumulators, modulating endophytes to resist/adapt to such environmental conditions. It is also possible that the metal hyperaccumulating plants may simultaneously be colonized by different metal-resistant endophytic bacteria ranging wide variety of gram-positive and gram-negative bacteria (Rajkumar et al., 2009).

However, the success of phytoremediation of metals depends upon a plant's ability to tolerate to accumulate high concentrations of the metals, while yielding a large plant biomass (Grčman et al., 2001). Due to their importance for practical applications, metal-tolerant plant– microbe associations have been the objective of particular attention due to the potential of microorganisms for bioaccumulating metals from polluted environment or its effects on metal mobilization/immobilization and consequently enhancing metal uptake and plant growth. Synergistic use of plants and microbes has been profitable for cleanup of metalliferous soils (Jing et al., 2007; Glick, 2010).

This chapter describes how the beneficial partnerships between plants and their associated bacteria can be exploited as a strategy to accelerate plant biomass production and influence plant metal accumulation or stabilization with better performance abilities such as adaptive strategies, metal mobilization, and immobilization mechanisms.

Role of metal-resistant bacteria on plant growth in metal-contaminated soils

In both natural and managed ecosystems, plant-associated bacteria play a key role in host adaptation to a changing environment. These microorganisms can alter plant cell metabolism, so that upon exposure to heavy metal stress, the plants are able to tolerate high concentrations of metals and thus can better withstand the challenge (Welbaum et al.,

2004). Several of the plant-associated bacteria have been reported to accelerate phytoremediation in metal-contaminated soils by promoting plant growth and health and play a significant role in accelerating phytoremediation (Compant et al., 2010; Dary et al., 2010).

(a) Plant growth promoting factors

The phytohormone ethylene (C2H4) has a central role in modulating the growth and cellular metabolism of plants and been believed to be involved in disease-resistant biotic/abiotic stress tolerance, plant–microbe partnership, and plant nutrient cycle. Among its key role in inducing various physiological changes in plants at molecular level, the overproduction of ethylene can cause the inhibition of root elongation, lateral root growth, and root hair formation; however, bacteria are capable of alleviating the stress-mediated impact on plants by enzymatic hydrolysis of 1 aminocyclopropane-1-carboxylic acid (ACC) . ACC is involved in biosynthetic pathway of ethylene, as an intermediate in the conversion of methionine to ethylene following biosynthetic sequence: methionine–S-adenosylmethionine (SAM)–ACC–C2H4 (Adams and Yang, 1979). In general, ACC is exuded from plant roots or seeds and then taken up by the ACC-utilizing bacteria before its oxidation by the plant ACC oxidase and cleaved by ACC deaminase to α-ketobutyrate (αKB) and ammonia. The bacteria utilize the ammonia evolved from ACC as a sole nitrogen source and thereby decrease ACC within the plant   with the concomitant reduction of plant ethylene. The decreased ethylene levels in plants hosting ACC-utilizing bacteria derive benefit by stress alleviation and enhanced plant productivity. Since SAM is converted by ACC synthase to ACC, the ACC synthase protein seems to play a main controlling role in ethylene biosynthesis pathway. In the absence of ACC-utilizing bacteria, ACC is oxidized by ACC oxidase to form ethylene, cyanide, and CO2. Moreover, the multigene family of ACC synthase and ACC oxidase is regulated independently by biotic and abiotic factors, thus possibly influence the plant ethylene biosynthesis pathway.

An unknown phosphatase (Ptase) or other mechanism regulates the turnover of the ACC synthase protein from the phosphorylated form (Pi) to the less stable non-phosphorylated form (Hardoim et al., 2008).

Phytohormones that are produced by plant-associated bacteria, including indole-3-acetic acid (IAA), cytokinins, and gibberellins, can frequently stimulate germination, growth, reproduction, and protect plants against both biotic and abiotic stress. As the most studied phytohormones, IAA produced in the plant shoot and transported basipetally to the root tips associated with cell elongation and cell division contributes to plant growth and plant defense system development. Further, plant–microbe interactions were determined by different IAA biosynthesis pathways. For instance, the beneficial plant-associated bacteria synthesize IAA via the indole-3-pyruvate (IPyA) pathway, whereas pathogenic bacteria mainly use the indole-3-acetamide (IAM) pathway (Hardoim et al., 2008)

In addition, the PGPB may also contribute in reducing the metal phytotoxicity by biosorption and bioaccumulation mechanisms. Since the bacterial cells (approximately 1.0–1.5 mm3) have an extremely high ratio of surface area to volume, they could adsorb a greater amount of heavy metals than inorganic soil components (e.g., kaolinite, vermiculite) either by a metabolism-independent passive, or by a metabolism-dependent active process (Khan et al., 2007). Several authors have pointed out that bacterial biosorption/bioaccumulation mechanism, together with other plant growth promoting features including the production of ACC deaminase and phytohormones, accounted for improved plant growth in metal-contaminated soils (Kumar et al., 2009).

B) Micro- and macro-nutrient provider

Iron is a necessary cofactor for many enzymatic reactions and hence is an essential nutrient for virtually all organisms. In the aerobic conditions, iron exists predominantly as ferric state (Fe^{3+}) and reacts to form highly insoluble hydroxides and oxyhydroxides that are largely unavailable to plants and microorganisms. To acquire sufficient iron, siderophores produced by bacteria can bind Fe^{3+} with a high affinity to solubilize this metal for its

LAP LAMBERT ACADEMIC PUBLISHING AG & CO. KG, DUDWELLER LANDSTR, GERMANY

efficient uptake. Although strategy II plants (Poaceae) release phytosiderophores to enhance their Fe uptake, phytosiderophores typically have a lower affinity for iron than microbial siderophores. Thus, these plants are unable to uptake sufficient amounts of iron. Further, heavy metals that are accumulated in excess in plant tissues can cause changes in various vital growth processes and have negative effects on iron nutrition. Under such conditions, the siderophore producing rhizosphere bacteria might offer a biological rescue system that is capable of chelating $Fe3+$ and making it available to plant roots. The roots could then take up iron from siderophores–Fe complexes possibly via the mechanisms such as chelate degradation and release of iron, the direct uptake of siderophore–Fe complexes, and/or a ligand exchange reaction (Rajkumar et al., 2010). Several examples of increased Fe uptake in plants with concurrent stimulation of plant growth as a result of PGPB inoculations have been reported (Barzanti et al., 2007). Siderophores also promote bacterial IAA synthesis by reducing the detrimental effects of heavy metals through chelation reaction (Dimkpa et al., 2008a).

Phosphorus (P) is a major essential macronutrient for biological growth and development. Soluble P is often the limiting mineral nutrient for biomass production in natural ecosystems only taken up in monobasic ($H2PO4$ −) or dibasic ($HPO42−$) soluble forms (Glass, 1989), and the elevated levels of heavy metals in soil interfere with P uptake and lead to plant growth retardation.

Under metal stressed conditions, most metal-resistant PGPB can either convert these insoluble phosphates into available forms through acidification, chelation, exchange reactions, and release of organic acids or mineralize organic phosphates by secreting extracellular phosphatases. An increase in P availability to plants through the inoculation of phosphate-solubilizing bacteria has been reported in pot experiments and under field conditions. In addition, fixation of atmospheric nitrogen is a metabolic virtuosity of endophytes and rhizobacteria and colonization offers benefit to the host (Dobbelaere et al., 2003).

Role of metal-resistant bacteria on metal accumulation by plants

Although several conditions, for example, the plant growth, metal tolerance/accumulation, bacterial colonization, and plant growth promoting potentials must be met for microbe assisted phytoremediation to become effective, the concentration of bioavailable metals in the rhizosphere greatly influences the quantity of metal accumulation in plants, because a large proportion of heavy metals are generally bound to various organic and inorganic constituents in polluted soil and their phytoavailability is closely related to their chemical speciation (McBride, 1994). The metabolites released by PGPB (e.g., siderophores, biosurfactants, organic acids, plant growth regulators, etc.) can alter the uptake of heavy metals indirectly and directly: indirectly, through their effects on plant growth dynamics, and directly, through acidification, chelation, precipitation, immobilization, and oxidation–reduction reactions in the rhizosphere.

(a) Microbial-induced metal mobilization in phytoextraction

Plant-associated bacteria can potentially improve phytoextraction by altering the solubility, availability, and transport of heavy metal and nutrients by reducing soil pH, release of chelators, P solubilization, or redox changes. Among the various metabolites produced by PGPB, the siderophores play a significant role in metal mobilization and accumulation (Dimkpa, et al., 2009b; Rajkumar et al., 2010), as these compounds produced by PGPB solubilize unavailable forms of heavy metal-bearing Fe but also form complexes with bivalent heavy metal ions that can be assimilated by root mediated processes (Baraud et al., 2009) Recently, Braud et al. (2009) investigated the release of Cr and Pb in soil solution after inoculation of various PGPB and found that the siderophores producing PGPB, Pseudomonas aeruginosa was able to solubilize large amounts of Cr and Pb in soils solution. As an opportunistic pathogen, P. aeruginosa used in these experiments are therefore only a model system since regulatory agencies will never give permission for the deliberate release of this bacterium to the environment. Furthermore, these authors reported that inoculation of Zea mays with P. aeruginosa increased Cr and Pb uptake into the shoots

LAP LAMBERT ACADEMIC PUBLISHING AG & CO. KG, DUDWELLER LANDSTR, GERMANY

by a factor of 4.3 and 3.4, respectively. Similarly, the role of siderophores produced by Streptomyces tendae F4 in Cd uptake by bacteria and sunflower plant was investigated (Dimkpa et al., 2009b). Bacterial culture filtrates containing hydroxamate siderophores secreted by S. tendae F4 significantly enhanced uptake of Cd by the plant, compared to the control. This study showed that siderophores can help to reduce metal toxicity in bacteria while simultaneously facilitating the uptake of such metals by plants. In another study, these effects of siderophores were also reported by Dimkpa et al. (2009a), who found that the addition of siderophore-containing culture filtrate of S. tendae F4 to metal-contaminated soils increased Cd and Cr uptake by cowpea. These studies highlighted the potential of inoculating soils or plants with siderophore producing PGPB to further improve their phytoextraction efficiency.

In addition, certain PGPB have been shown to increase heavy metal mobilization by the secretion of low-molecular-mass organic acids comprising gluconate, 2-ketogluconate, oxalate, citrate, acetate, malate, and succinate, etc. An example is the release of 5-ketogluconic acid by endophytic diazotroph Gluconacetobacter diazotrophicus, which dissolves various Zn sources such as ZnO, $ZnCO3$, or $Zn3 (PO4)2$, thus making Zn available for plant uptake (Saravanan et al., 2007). Although it is well accepted that organic acids produced by PGPB play an important role in the mobilization of heavy metals and mineral nutrients, the inoculation effects of organic acids producing bacteria on plant growth and metal accumulation in plants are still poorly understood. The biosurfactants produced by PGPB have also been demonstrated to enhance heavy metal mobilization in contaminated soils (Braud et al., 2006). For instance, a study showed that the inoculation of soils with biosurfactant producing Bacillus sp. J119 significantly enhanced biomass of tomato plants and Cd uptake in plant tissue. From these studies, it can be concluded that by inoculating the seeds/rhizosphere soils with selected metal mobilizing bacteria, it should be possible to improve bioavailable metal concentrations for plant uptake and thereby phytoextraction potential in metal-contaminated soils.

(b)Microbial-induced metal immobilization in phytostabilization

The use of plant-associated bacteria in phytostabilization strategies may assist plant growth and tolerance to metals, but can also reduce the metal uptake or translocation to aerial parts of plants by decreasing the metal bioavailability in the rooting medium. For survival under metal-stressed environment, plant-associated bacteria have evolved several mechanisms by which they can immobilize or transform metals rendering them inactive to tolerate the uptake of heavy metal ions. The mechanisms that are generally proposed for heavy metal resistance in bacteria are (1) exclusion of metal by a permeability barrier or by active export of metal from the cell; (2) intracellular physical sequestration of metal by binding extracellular polymers or extra cellular sequestration; (3) detoxification where metal is chemically modified to render it less active (Rouch et al.,1995) For instance, binding of metals to anionic functional groups (i.e., sulfhydryl, carboxyle, hydoxyle, salfonate, amine and amide groups) immobilizes the metal and prevents its entry into the plant root. Similarly, the metal binding extracellular polymers comprising polysaccharides, proteins, humic substances etc. may detoxify metals by chelating the heavy metals (Pulsawat et al., 2003). The bacterial siderophores and organic acids can also reduce the metal bioavailability and toxicity by chelating the metal ions (Tripathi et al., 2005; Dimkpa et al., 2008b). According to Dimkpa et al. (2008b), the decreasing Ni concentration in cowpea plants is indicative of a Nibinding potential of hydroxamate siderophores. Further, metalbiosorption by microbial inoculants is particularly interesting from a phytostabilization point of view. For instance, Madhaiyan et al. (2007) reported that inoculation with endophytic bacteria, Magnaporthe oryzae and Burkholderia sp. increased plant growth but reduced the Ni and Cd accumulation in roots and shoots of tomato and also their availability in soil. This effect was due to the increased metal biosorption and bioaccumulation by bacterial strains. In addition, bacteria can also interact directly with the heavy metals to reduce their toxicity and/or modulate their bioavailability: metal dissolution by bacterial production of strong acids (i.e., H2SO4 produced by Thiobacillus); production of ammonia or organic bases resulting in metal hydroxide precipitates; fixation of Fe andMn on the cell surface in the form of hydroxides or some other insoluble metal salts; biotransformation via methylation, demethylation, volatilization, complex formation, oxidation, or reduction (Chen and Cutright, 2003). Although the establishment of a successful vegetative cover on metal-contaminated soils is challenging, the beneficial bacteria immobilizing heavy metals and enhancing the plant tolerance to high metal concentrations and/or promoting plant growth could provide a practical tool for speeding up the phytostabilization process.

The role of soil microbiota, specifically rhizospheric and endophytic microorganisms, in the development of phytoremediation techniques has to be elucidated in order to speed up the process and to optimize the rate of mobilization/absorption/accumulation of pollutants. To efficiently phytoremediate metal-contaminated soils, the bioavailability of metals to plant roots is considered to be a critical requirement for plant metal bioconcentration or bioimmobilization to occur. In this regard, it may be possible to employ beneficial bacteria to alter the bioavailability of metals for improving phytoremediation of metal contaminants on large scale in the environment. Based on the foregoing account, *supra vide*, microbe assisted phytoremediation is a reliable and dependable process.

Plant–endophyte partnerships in phytoremediation

Plant uptake of organic contaminants

Plant uptake is the first crucial step in whole plant metabolism of organics. In case of constant plant and environmental features, the lipophilicity of the compound – expressed as its octanol–water partition coefficient (Kow) – was shown to be the determining factor for root entry and translocation. Organic contaminants with a log Kow < 1 are considered to be very water-soluble, and plant roots do not generally accumulate them at a rate surpassing passive influx into the transpiration stream (Cunningham et al., 1993). Contaminants with a log Kow > 3.5 show high sorption to the roots but slow or no translocation to the stems and leaves (Trapp et al., 2001). However, plants readily take up organic contaminants with a log Kow between 0.5 and 3.5, as well as weak electrolytes (weak acids and bases or amphoteres as herbicides).

Plant-bacteria synergism for the phytoremediation of organics

After plant uptake, the organic compound may be metabolized and/or released into the atmosphere via evapotranspiration through the stem and/or leaves. Although plants often metabolize or sequester organics, they are at a significant disadvantage in two ways (Burken et al., 2003):

(1) Being photoautotrophic, plants do not rely on organic molecules as a source of energy or carbon. By consequence, unlike microorganisms, during evolution plants were not under selective pressure to develop the capacity to degrade chemically recalcitrant molecules, leading to a much more limited spectrum of chemical structures that they can metabolize;

(2) to avoid build-up and potential toxicity to sensitive organelles, plant metabolism of organic molecules (other than photosynthates) consists of general transformations to more water-soluble forms, and sequestration processes (green-liver model: Burken et al., 2003). By contrast, microbial metabolism often ends with the organics being converted into CO_2, water and cellular biomass. Therefore, in order to obtain a more efficient degradation of organic compounds, plants depend on their associated microorganisms. Plants themselves have a positive effect on the microbial degradation of organic contaminants (Crowley et al., 1997). This increased degradation potential is the result of higher microbial densities and metabolic activities in the rhizosphere due to microbial growth on root exudates and cell debris originating from the plant roots. Moreover, dense populations of diverse heterotrophic microorganisms are living in the rhizosphere, the phyllosphere and inside the plant (endophytes). These microbial associations increase the capacity for a stepwise transformation of organic contaminants by consortia and provide a habitat that is conducive to genetic exchange and gene rearrangements. The emerging picture suggests that plants draw pollutants, including PAHs; into their rhizosphere to varying extents via the transpiration stream (Harvey et al., 2002) Subsequent degradation can occur in the plant

itself, or in the rhizosphere, or both. However, compounds with a log Kow between 0.5 and 3.5 seem to enter the xylem faster than the soil, and rhizosphere microflora can degrade them, even if the latter is enriched with bacteria capable of degrading the compound (Trapp et al., 2000). Therefore, after this class of compounds is taken up by the plants, endophytes seem to be especially suitable for the degradation of these compounds.

Although successfully applied in several demonstration projects, large-scale field application of phytoremediation of organic pollutants is limited by several restrictions:

(a) The levels of contaminants tolerated by the plant,

(b) The often limited bioavailability of the contaminants and,

(c) In certain cases, unacceptable levels of evapotranspiration of volatile organic contaminants to the atmosphere. A possible solution to conquer these constraints is the use of genetically manipulated plants specifically tailored for phytoremediation purposes (Doty 2008). However, since bacteria are much easier to manipulate than plants, and natural gene transfer between closely related environmental and endophytic species is possible (avoiding the limitations of using GMOs), many studies have focused on the use of natural or engineered plant-associated bacteria. The state of the art of rhizosphere 'engineering' for accelerated rhizodegradation of persistent organic contaminants was recently reviewed in (Dzantor 2007). Even when an efficient rhizodegradation seems possible, compounds with lipophilicity in the optimum range seem to enter the root xylem before the soil and rhizosphere microflora can degrade them. Since the residence time of contaminants in the xylem ranges from several hours to up to two days (Mc Crady 1987) (engineered) degrading endophytes colonizing the xylem are perfect candidates to reduce phytotoxicity and to avoid evapotranspiration of contaminants or their degradation intermediates into the environment. If no naturally occurring endophytes with the desired metabolic properties are available, endophytic bacteria can be isolated, equipped with the appropriate degradation pathways and subsequently reinoculated in their host plant. The general idea behind the use of engineered endophytes to improve phytoremediation is to complement the metabolic properties of their host plant. Proof of this concept was provided by inoculating yellow

LAP LAMBERT ACADEMIC PUBLISHING AG & CO. KG, DUDWELLER LANDSTR, GERMANY

lupine plants (Barac et al., 2009) and poplar (Taghavi et al., 2005) with endophytic bacteria able to degrade toluene, which resulted in decreased toluene phytotoxicity and significantly lowered toluene evapotranspiration.

As many catabolic pathways are occurring in soil bacteria, where they are often encoded on self-transferable plasmids or transposons, natural gene transfer offers huge potential for the a la carte construction of endophytic bacteria with appropriate catabolic pathways. Heterologous expression of these catabolic functions might not constitute a major problem, especially when the donor and the recipient endophytic strains are closely related. Other applications than remediation can also be envisaged, such as protection of the food chain by reducing residual levels of agrochemicals in food crops. Recently, the use of bacterial endophytes for reducing levels of toxic herbicide residues in crop plants was successfully demonstrated (Germaine et al., 2006). Inoculation of pea plants (*Pisum sativum*) with a poplar endophyte able to degrade 2,4 dichlorophenoxyacetic acid (2,4- D) resulted in an increased ability to remove 2,4-D from the soil, while the plants did not accumulate 2,4-D in the tissues nor showed toxic effects (Germaine et al., 2006).

Although it is obvious that the application of engineered plant-associated bacteria to improve phytoremediation of organic contaminants has high potential, an important issue is the persistence and the stability of the engineered organisms and their degradation capabilities in association with plants growing in the field. As long as there is a selection pressure, there will be a selective advantage for those community members possessing the appropriate degradation characteristics. However, instead of integrating a new strain, the endogenous microbial community can also get adapted through horizontal gene transfer. Horizontal gene transfer has been illustrated to perform an important role in the adaptation of microbial communities to environmental stress factors, including rhizospheric (Ronchel et al., 2000; Devers et al., 2005) and endophytic communities (Taghavi et al., 2005). This may have the practical advantage that no long-term establishment of inoculants is required.

Endophytes take the challenge to improve phytoextraction of toxic metals

Metal availability, metal uptake and phytotoxicity for the plant are the main limiting factors for the application of phytoextraction. Phytoextraction is a long-term process, it may not be able to remove 100% of the contamination and, until now, its efficiency has only been demonstrated for some metals (Chaney et al., 2007). To optimize phytoextraction, genetic manipulation of plants as well as manipulation of the plant associated microbial communities has been considered .Possible manipulation strategies of the plant associated community to improve the efficiency of phytoextraction include (a) isolation of associated bacteria, followed by equipping them with (a1) metabolic pathways for the synthesis of natural chelators, such as citric acid to improve metal availability for plant uptake and translocation and with (a2) metal sequestration systems to reduce phytotoxicity; and re-inoculation of these modified bacteria (Valls et al., 2002)as well as (b) enrichment of bacteria present in planta (Li 2007). For instance, *Lupinus luteus* L, when grown on a nickel enriched substrate and inoculated with the engineered nickel-resistant endophytic bacterium B. cepacia L.S.2.4::ncc-nre, showed a significant increase (30%) of nickel concentration in the roots, whereas the nickel concentration in the shoots remained comparable with that of the control plants (Lodewyckx 2001).

Phytoremediation of mixed waste pollution

Although there exists an obvious difference in phytoremediation potential whether organics or metals are the primary targets; at most contaminated sites, plants and their associated microorganisms will have to deal with mixed contamination. Remediation of these mixed waste sites is generally intricate. The occurrence of toxic metals potentially inhibits a broad range of microbial processes, including the degradation of organic pollutants (Sandrin et al., 2003) A very promising strategy to tackle the mixed waste situation is the use of endophytes that are capable of (a) degrading organic contaminants and of (b) dealing with, or in the ideal scenario, accelerating the extraction of toxic metals. It has been shown that engineering of rhizobacteria for TCE degradation and heavy metal (Cd) accumulation resulted in an increased Cd accumulation but also in a lowered toxic effect of Cd on the TCE degradation (Lee et al., 2006). Similar improvements are expected when these engineered rhizobacteria are inoculated onto plant roots.

References

1. Mansour SA, Gad MF. 2010. Risk assessment of pesticides and heavy metals contaminants in vegetables: a novel bioassay method using Daphnia magna Straus. Food Chem Toxicol; 48:377–89.

2. Lim SR, Schoenung JM. 2010. Human health and ecological toxicity potentials due to heavy metal content in waste electronic devices with flat panel displays. J Hazard Mater; 177:251–9.

3. Glick BR. 2010. Using soil bacteria to facilitate phytoremediation. Biotechnol Adv; 28: 367–74.

4. Baysse, C., DeVos, D., Naudet, Y., Vander monde, A., Ochsner, U., Meyer, J.M., Budzikiewicz, H., Schäfer, M., Fuchs, R., Cornelis, P., 2000. Vanadium interferes with siderophore-mediated iron uptake in Pseudomonas aeruginosa. Microbiology 146, 2425–2434.

5. Yoshihara, T., Hodoshima, H., Miyano, Y., Shoji, K., Shimada, H., Goto, F., 2006. Cadmium inducible Fe deficiency responses observed from macro and molecular views in tobacco plants. Plant Cell Rep. 25, 365–373.

6. Christian, O.D., Aleš, S., Paulina, D., Andre, S., Wilhelm, B., Erika, K., 2008. Involvement of siderophores in the reduction of metal-induced inhibition of auxin synthesis in Streptomyces spp. Chemosphere 74, 19–25.

7. Nogaw, K., Kido, T., 1996. Ital–Ital disease and health effects of cadmium. In: Chang, L.W. (Ed.), Toxicology of Metals. CRC Press, Boca Raton, FL, pp. 353–369

8. Xiao, X., Luo, S.L., Zeng, G.M., Wei, W.Z., Wan, Y., Chen, L., Guo, H.J., Cao, Z., Yang, L.X., Chen, J.L., Xi, Q., 2010. Biosorption of cadmium by endophytic

fungus (EF) Microsphaeropsis sp. LSE10 isolated from cadmium hyperaccumulator Solanum nigrum L. Bioresour. Technol. 101, 1668–1674.

9. Chen, G.Q., Zeng, G.M., Tang, L., Du, C.Y., Jiang, X.Y., Huang, G.H., Liu, H.L., Shen, G.L., 2008. Cadmium removal from simulated wastewater to biomass byproduct of Lentinus edodes. Bioresour. Technol. 99, 7034–7040.

10. Tang, L., Zeng, G.M., Shen, G.L., Li, Y.P., Zhang, Y., Huang, D.L., 2008. Rapid detection of picloram in agricultural field samples using a disposable immunomembrane-based electrochemical sensor. Environ. Sci. Technol. 42, 207–1212.

11. Wu CH, Wood TK, Mulchandani A, Chen W. 2006a. Engineering plant–microbe symbiosis for rhizoremediation of heavy metals. Appl Environ Microbiol; 72:1129–34.

12. Grčman H, Velikonja-Bolta S, Vodnik D, Kos B, Leštan D. EDTA enhanced heavy metal phytoextraction: metal accumulation, leaching, and toxicity. Plant Soil 2001; 235: 105–14.

13. Jing Y, He Z, Yang X. Role of soil rhizobacteria in phytoremediation of heavy metal contaminated soils. J Zhejiang Univ Sci. 2007; 8:192–207.

14. Cheng, S., 2003. Heavy metal pollution in China: origin, pattern and control. Environ. Sci. Pollut. R. 10, 192–198.

15. Turgut, C., 2003. The contamination with organochlorine pesticides and heavy metals in surface water in Kucuk menderes river in Turkey, 2000–2002. Environ. Int. 29, 29–32.

16. Alloway, B.J., 1995. Heavy Metal in Soils. second ed. Chapman and Hall, London

17. Diels, L., van der Lelie, N., Bastiaens, L., 2002. New development in treatment of heavy metal contaminated soils. Rev. Environ. Sci. Biotechnol. 1, 75–82.

18. Holliday P (1989). A Dictionary of Plant Pathology. Cambridge University Press, Cambridge.

LAP LAMBERT ACADEMIC PUBLISHING AG & CO. KG, DUDWELLER LANDSTR, GERMANY

19. Schulz B & Boyle C 2006. What are endophytes? Microbial Root Endophytes (Schulz BJE, Boyle CJC & Sieber TN, eds), pp. 1–13. Springer-Verlag, Berlin.

20. Strobel G, Daisy B, Castillo U & Harper J 2004. Natural products from endophytic microorganisms. J Nat Prod 67: 257–268.

21. Van Aken, B. et al. (2004a) Methylobacterium populum sp. nov.,a novel aerobic, pink-pigmented, facultatively methylotrophic, methane-utilizing bacterium isolated from poplar trees (*Populus deltoides x nigra* DN34). Int. J. Syst. Evol. Microbiol. 54, 1191–1196

22. Van Aken, B. et al. (2004b) Biodegradation of Nitro-Substituted Explosives 2,4,6-Trinitrotoluene, Hexahydro-1,3,5-Trinitro-1,3,5- Triazine, and Octahydro-1,3,5,7-Tetranitro-1,3,5-Tetrazocine by a Phytosymbiotic Methylobacterium sp. Associated with Poplar Tissues (*Populus deltoides!nigra* DN34). Appl. Environ. Microbiol. 70, 508–517

23. Barzanti R, Ozino F, Bazzicalupo M, Gabbrielli R, Galardi F, Gonnelli C, Isolation and characterization of endophytic bacteria from the nickel hyperaccumulator plant *Alyssum bertolonii*. Microb Ecol 2007; 53:306–16.

24. Mastretta C, Taghavi S, van der Lelie D,Mengoni A, Galardi F, Gonnelli C,. Endophytic bacteria from seeds of *Nicotiana tabacum* can reduce cadmium phytotoxicity. Int J Phytorem 2009; 11:251–67.

25. Ryan RP, Germaine K, Franks A, Ryan DJ, Dowling DN. 2008. Bacterial endophytes: recent developments and applications. FEMS Microbiol Lett; 278:1–9.

26. Compant S, Reiter B, Sessitsch A, Nowak J, Clement C, Barka EA. 2005. Endophytic colonization of *Vitis vinifera* L. by a plant growth-promoting bacterium, *Burkholderia sp.* strain PsJN. Appl Environ Microbiol; 71:1685–93.

27. Idris R, Trifonova R, Puschenreiter M, Wenzel WW, Sessitsch A. 2004. Bacterial communities associated with flowering plants of the Ni hyperaccumulator *Thaspi goesingense*. Appl Environ Microbiol; 70:2667–77

LAP LAMBERT ACADEMIC PUBLISHING AG & CO. KG, DUDWELLER LANDSTR, GERMANY

28. Rajkumar M, Ae N, Freitas H. 2009. Endophytic bacteria and their potential to enhance heavy metal phytoextraction. Chemosphere; 77:153–60.

29. Welbaum GE, Sturz AV, Dong Z, Nowak J. 2004. Managing soil microorganisms to improve productivity of agro-ecosystems. Crit Rev Plant Sci; 23:175–93.

30. Compant S, Clément C, Sessitsch A. 2010. Plant growth-promoting bacteria in the rhizo- and endosphere of plants: their role, colonization, mechanisms involved and prospects for utilization. Soil Biol Biochem; 42:669–78.

31. Dary M, Chamber-Pérez MA, Palomares AJ, Pajuelo E. 2010. "In situ" phytostabilisation of heavy metal polluted soils using *Lupinus luteus* inoculated with metal resistant plant-growth promoting rhizobacteria. J Hazard Mater; 177:323–30.

32. Adams D. O, Yang S. F. 1979. Ethylene biosynthesis: identification of l-amino cyclopropane carboxylic acid as an intermediate in the conversion of methionine to ethylene. Proc Natl Acad Sci U S A; 76:170–4.

33. Hardoim PR, van Overbeek LS, van Elsas JD. 2008. Properties of bacterial endophytes and their proposed role in plant growth. Trends Microbiol; 16:463–71.

34. Khan MS, Zaidi A, Wani PA. 2007. Role of phosphate-solubilizing microorganisms in sustainable agriculture—a review. Agron Sustain Dev; 27:29–43.

35. Kumar KV, Srivastava S, Singh N, Behl HM. 2009. Role of metal resistant plant growth promoting bacteria in ameliorating fly ash to the growth of *Brassica juncea*. J Hazard Mater; 170:51–7.

36. Barzanti R, Ozino F, Bazzicalupo M, Gabbrielli R, Galardi F, Gonnelli C, 2007. Isolation and characterization of endophytic bacteria from the nickel hyperaccumulator plant Alyssum bertolonii. Microb Ecol; 53:306–16.

37. Dimkpa CO, Svatoš A, Dabrowska P, Schmidt A, Boland W, Kothe E. 2008a. Involvement of siderophores in the reduction of metal-induced inhibition f auxin synthesis in Streptomyces spp. Chemosphere 74:19–25.

LAP LAMBERT ACADEMIC PUBLISHING AG & CO. KG, DUDWELLER LANDSTR, GERMANY

38. Glass ADM. 1989. Plant Nutrition: An Introduction to Current Concepts. Boston: Jones and Bartlett Publishers;. p. 234.

39. Dobbelaere S, Vanderleyden J, Okon Y. 2003. Plant growthpromoting effects of diazotrophs in the rhizosphere. Crit Rev Plant Sci; 22:107–49.

40. Braud A, Jézéquel K, Bazot S, Lebeau T. 2009. Enhancedphytoextraction of an agricultural Cr-,Hgand Pb-contaminated soil by bioaugmentation with siderophore producing bacteria. Chemosphere; 74:280–6.

41. McBride MB. 1994. Environmental Chemistry in Soils. Oxford: Oxford Univ. Press. p. 406.

42. Dimkpa CO, Merten D, Svatoš A, Büchel G, Kothe E. 2009a. Metal-induced oxidative stress impacting plant growth in contaminated soil is alleviated by microbial siderophores. Soil Biol Biochem; 41:154–62.

43. Dimkpa CO, Merten D, Svatos A, Büchel G, Kothe E. 2009b. Siderophores mediate reduced and increased uptake of cadmium by *Streptomyces tendae* F4 and sunflower (*Helianthus annuus*), respectively. J Appl Microbiol; 107:1687–96.

44. Braud A, Jézéquel K, Vieille E, Tritter A, Lebeau T. 2006. Changes in extractability of Cr and Pb in a polycontaminated soil after bioaugmentation with microbial producers of biosurfactants, organic acids and siderophores. Water Air Soil Pollut; 6:3–4.

45. Saravanan VS, Madhaiyan M, Thangaraju M. 2007. Solubilization of zinc compounds by the diazotrophic, plant growth promoting bacterium *Gluconacetobacter diazotrophicus*. Chemosphere; 66:1794–8.

46. Rouch DA, Lee TOB, Morby AP. 1995. Understanding cellular responses to toxic agents: a model for mechanisms-choice in bacterial resistance. J Ind Microbiol; 14: 132–41.

47. PulsawatW, Leksawasdi N, Rogers PL, Foster LJR. 2003. Anions effects on biosorption ofMn(II) by extracellular polymeric substance (EPS) from *Rhizobium etli*. Biotechnol Lett; 25: 1267–70.

48. Tripathi M, Munot HP, Shouche Y, Meyer JM, 2005. Goel R. Isolation and functional characterization of siderophore-producing lead- and cadmium-resistant *Pseudomonas putida* KNP9. Curr Microbiol; 50:233–7.

49. Madhaiyan M, Poonguzhali S, Sa T. 2007. Metal tolerating methylotrophic bacteria reduces nickel and cadmium toxicity and promotes plant growth of tomato (*Lycopersicon esculentum* L.). Chemosphere; 69:220–8.

50. Chen H, Cutright TJ. 2003. Preliminary evaluation of microbial mediated precipitation of cadmium, chromium, and nickel by rhizosphere consortium. J Environ Eng; 129:4–9

51. Cunningham SD, Berti WB: 1993. Remediation of contaminated soils with green plants: an overview. In Vitro Cellular Developmental Biol, 29:207-212.

52. Trapp S, Ko¨ hler A, Larsen LC, Zambrano KC, Karlson U: 2001. Phytotoxicity of fresh and weathered diesel and gasoline to willow and poplar trees. J Soils Sed, 1:71-76.

53. Burken JG: 2003. Uptake and metabolism of organic compounds: green-liver model. In Phytoremediation: Transformation and Control of Contaminants. Edited by McCutcheon SC, Schnoor JL.:59-84.

54. Crowley DE, Alvey S, Gilbert ES: 1997. Rhizosphere ecology of xenobiotic degrading microorganisms. In Phytoremediation of Soil and Water Contaminants. Edited by Kruger EL, Anderson TA, Coats JR.

55. Harvey PJ, Campanella BF, Castro PM, Harms H, Lichtfouse E, Scha¨ ffner AR, Smrcek S, Werck-Reichhart D: 2002. Phytoremediation of polyaromatic hydrocarbons, anilines and phenols. Environ Sci Pollut Res Int, 9:29-47.

56. Trapp S, Zambrano KC, Kusk KC, Karlson U: 2000. A phytotoxicity test using transpiration of willows. Arch Environ Contam Toxicol, 39:154-160

57. Doty SL: Enhancing phytoremediation through the use of transgenics and endophytes. New Phytologist 2008, 179(2):318-333.

LAP LAMBERT ACADEMIC PUBLISHING AG & CO. KG, DUDWELLER LANDSTR, GERMANY

58. Dzantor EK: Phytoremediation: the state of rhizosphere 'engineering' for accelerated rhizodegradation of xenobiotic contaminants. J Chem Technol Biotech 2007, 82:228-232.

59. Mc Crady J, Mc Farlane C, Lindstrom F: The transport and affinity of substituted benzenes in soybean stems. J Exp Botany 1987, 38:1875-1890.

60. Taghavi S, Barac T, Greenberg B, Borremans B, Vangronsveld J, van der Lelie D: Horizontal gene transfer to endogenous endophytic bacteria from Poplar improves phytoremedeiation of toluene Appl Environ Microbiol 2005, 71:8500-8505.

61. Lee W, Wood TK, Chen W: Engineering TCE-degrading rhizobacteria for heavy metal accumulation and enhanced TCE degradation. Biotechnol Bioeng 2006, 95:399-403.

62. Sandrin et al., 2003 Sandrin TR, Maier RM: Impact of metals on the biodegradation of organic pollutants. Environ Health Perspect 2003, 111:1093-1101

63. Lodewyckx 2001 Lodewyckx C, Taghavi S, Mergeay M, Vangronsveld J, Clijsters H, van der Lelie D: The effect of recombinant heavy metal resistant endophytic bacteria on heavy metal uptake by their host plant. Int J Phytorem 2001, 3:173-187.

64. Li 2007 LiWC, Ye ZH, Wong MH: Effects of bacteria on enhanced metal uptake of the Cd/Zn-hyperaccumulating plant, Sedum alfredii. J Exp Bot 2007, 58:4173-4182.

65. Valls et al., 2002 Valls M, de Lorenzo V: Exploiting the genetic and biochemical capacities of bacteria for the remediation of heavy metal pollution. FEMS Microbiol rev 2002, 26:327-338

Druck: KN Digital Printforce GmbH · Schockenriedstraße 37 · 70565 Stuttgart